Common Core State Standards Station Activities for Mathematics II

1 2 3 4 5 6 7 8 9 10
ISBN 978-0-8251-7409-4

J. Weston Walch, Publisher
Portland, ME 04103
www.walch.com
Printed in the United States of America

Table of Contents

Introduction

Common Core State Standards Station Activities for Mathematics II is a collection of station-based activities to provide students with opportunities to practice and apply the mathematical skills and concepts they are learning in their Math II class. This book is organized to reflect the conceptual categories and domains found in the Common Core State Standards for Mathematics. It contains several sets of activities for each of the following Common Core high school mathematics conceptual categories and domains: Number and Quantity; Algebra; Functions; Geometry; and Statistics and Probability. You may use these activities as a complement to your regular lessons, or in place of direct instruction if formative assessment suggests students have the basic concepts but need practice. The debriefing discussions after each set of activities provide an important opportunity to help students reflect on their experiences and synthesize their thinking. The debrief also provides an additional opportunity for ongoing, informal assessment to inform instructional planning.

Implementation Guide

The following guidelines will help you prepare for and use the activity sets in this book.

Setting Up the Stations

Each activity set consists of four stations. Set up each station at a desk, or at several desks pushed together, with enough chairs for a small group of students. Place a card with the number of the station on the desk. Each station should also contain the materials specified in the teacher's notes, and a stack of student activity sheets (one copy per student). Place the required materials (as listed) at each station.

When a group of students arrives at a station, each student should take one of the activity sheets to record the group's work. Although students should work together to develop one set of answers for the entire group, each student should record the answers on his or her own activity sheet. This helps keep students engaged in the activity and gives each student a record of the activity for future reference.

Forming Groups of Students

All activity sets consist of four stations. You might divide the class into four groups by having students count off from 1 to 4. If you have a large class and want to have students working in small groups, you might set up two identical sets of stations, labeled A and B. In this way, the class can be divided into eight groups, with each group of students rotating through the "A" stations or "B" stations.

Assigning Roles to Students

Students often work most productively in groups when each student has an assigned role. You may want to assign roles to students when they are assigned to groups and change the roles occasionally. Some possible roles are as follows:

- Reader—reads the steps of the activity aloud
- Facilitator—makes sure that each student in the group has a chance to speak and pose questions; also makes sure that each student agrees on each answer before it is written down
- Materials Manager—handles the materials at the station and makes sure the materials are put back in place at the end of the activity
- Timekeeper—tracks the group's progress to ensure that the activity is completed in the allotted time
- Spokesperson—speaks for the group during the debriefing session after the activities

Timing the Activities

The activities in this book are designed to take approximately 10 minutes per station. Therefore, you might plan on having groups change stations every 10 minutes, with a 2-minute interval for moving from one station to the next. It is helpful to give students a "5-minute warning" before it is time to change stations. Since each activity set consists of four stations, the above time frame means that it will take about 50 minutes for groups to work through all stations.

Guidelines for Students

Before starting the first activity set, you may want to review the following "ground rules" with students. You might also post the rules in the classroom.

- All students in a group should agree on each answer before it is written down. If there is a disagreement within the group, discuss it with one another.
- You can ask your teacher a question only if everyone in the group has the same question.
- If you finish early, work together to write problems of your own that are similar to the ones on the student activity sheet.
- Leave the station exactly as you found it. All materials should be in the same place and in the same condition as when you arrived.

Debriefing the Activities

After each group has rotated through every station, bring students together for a brief class discussion. At this time you might have the groups' spokespersons pose any questions they had about the activities. Before responding, ask if students in other groups encountered the same difficulty or if they have a response to the question. The class discussion is also a good time to reinforce the essential ideas of the activities. The questions that are provided in the teacher's notes for each activity set can serve as a guide to initiating this type of discussion.

You may want to collect the student activity sheets before beginning the class discussion. However, it can be beneficial to collect the sheets afterward so that students can refer to them during the discussion. This also gives students a chance to revisit and refine their work based on the debriefing session.

Guide to Common Core State Standards Annotation

As you use this book, you will come across annotation symbols included with the standards for several station activities. The following descriptions of these annotation symbols are verbatim from the Common Core State Standards Initiative website, at http://www.walch.com/CCSS/00002.

Symbol: ★

Denotes: Modeling Standards

The Standards for Mathematical Practice describe varieties of expertise that mathematics educators at all levels should seek to develop in their students. These practices rest on important "processes and proficiencies" with longstanding importance in mathematics education. Specific modeling standards appear throughout the high school standards indicated by a star symbol (★).

From http://www.walch.com/CCSS/00006

Symbol: (+)

Denotes: College and Career Readiness Standards

The evidence concerning college and career readiness shows clearly that the knowledge, skills, and practices important for readiness include a great deal of mathematics prior to the boundary defined by (+) symbols in these standards.

From http://www.walch.com/CCSS/00004

Standards Correlations

The standards correlations that follow support the implementation of the Common Core State Standards for Mathematics II. This book includes station activity sets for the Common Core conceptual categories of Number and Quantity, Algebra, Functions, Geometry, and Statistics and Probability. The table that follows provides a listing of the available station activities organized by standard.

The left column lists the standard codes. The first letter of the code represents the Common Core conceptual category. This letter is followed by a dash and the initials of the domain name, which is then followed by the standard number. The middle column lists the title of the station activity set that corresponds to the standard, and the right column lists the page number where the station activity set can be found. The full text of the Common Core State Standards for Mathematics may be accessed at http://www.walch.com/CCSS/00005.

(*continued*)

Standards Correlations

CCSS addressed	Set title	Page number
	Geometry: Congruence	
G–CO.9	Parallel Lines and Transversals	72
G–CO.11	Rhombi, Squares, Kites, and Trapezoids	84
G–CO.12 G–CO.13	Circumcenter, Incenter, Orthocenter, and Centroid	95
	Geometry: Similarity, Right Triangles, and Trigonometry	
G–SRT.1b G–SRT.2	Similarity and Scale Factor	109
G–SRT.6 G–SRT.7 G–SRT.8★	Sine, Cosine, and Tangent Ratios, and Angles of Elevation and Depression	122
	Geometry: Circles	
G–C.1 G–C.2 G–C.4 (+) G–C.5	Special Segments, Angle Measurements, and Equations of Circles	133
G–C.2 G–C.5	Circumference, Angles, Arcs, Chords, and Inscribed Angles	145
	Statistics and Probability	
S–CP.1★ S–CP.2★ S–CP.3★	Probability	159

Materials List

Class Sets

- calculators (standard and graphing)
- rulers

Station Sets

- algebra tiles (40 blue, 40 green, 40 red, and 40 yellow)
- compasses (at least 4)
- drinking straws
- fair coins (at least 2)
- marbles (1 each of red, green, yellow, and blue)
- markers (2 red, 1 black, 2 blue, and 2 green)
- notecards
- pieces of colored yarn (3 red and 3 blue)
- plastic coffee can lid
- protractors (at least 2)
- tape measure
- tracing paper
- uncooked spaghetti noodles

Ongoing Use

- colored pencils or pens
- graph paper
- index cards (prepared according to specifications in teacher notes for many of the station activities)
- number cubes
- pencils
- scissors
- tape
- white paper

Number and Quantity

Set 1: Operations with Complex Numbers

Instruction

Goal: To give students practice in adding, subtracting, multiplying, and dividing complex numbers; to help students recognize the relationship between complex numbers in radical form and numbers in $a + bi$ form

Common Core State Standards

N–CN.1 Know there is a complex number i such that $i^2 = -1$, and every complex number has the form $a + bi$ with a and b real.

N–CN.2 Use the relation $i^2 = -1$ and the commutative, associative, and distributive properties to add, subtract, and multiply complex numbers.

N–CN.3 (+) Find the conjugate of a complex number; use conjugates to find moduli and quotients of complex numbers.

Student Activities Overview and Answer Key

Station 1

Students race their partner to complete addition and subtraction problems involving complex numbers. Students check each other's work.

Answers

1. $3 + 8i$
2. $13 - 4i$
3. $22 + 5i$
4. $26 + 3i$
5. 5
6. $3a + 4gi$
7. $9i - 26$
8. 4
9. $7 - 7i$
10. $-4 + 17i$

Station 2

Students work with a partner to multiply complex numbers.

Answers

1. $-13 + 11i$
2. $-2 + 34i$
3. $1 + 8i$
4. $\frac{1}{2} + \frac{81}{4}i$
5. $-8 - 12i$

6. $13 - i$
7. $47 - 28i$
8. $50 + 10i$
9. 85

Station 3

Students work with groups to identify the conjugate c of complex numbers and solve division problems.

Answers

1. $c = 5 + 6i$

   $\frac{3}{61} + \frac{28}{61}i$
2. $c = 2 - i$

   $\frac{11}{5} + \frac{2}{5}i$
3. $c = 3 + 2i$

   $\frac{11}{13} + \frac{16}{13}i$
4. $c = 3 + 2i$

   $\frac{5}{13} + \frac{12i}{13}$
5. $c = 7 - 3i$

   1
6. $c = 2 - i$

   $\frac{13}{5} - \frac{14i}{5}$
7. $c = 5 - 3i$

   $\frac{12}{17} - \frac{14i}{17}$
8. $c = 4 - 2i$

   $\frac{1}{2} - \frac{i}{2}$

Station 4

Students work in groups to solve equations involving complex numbers, sometimes in radical form, sometimes in $a + bi$ form.

Answers

1. $8+\frac{i}{2}$
2. $3+4i$
3. $-2+3i$
4. $-20-20i$
5. $\frac{2}{3}-\frac{2i}{3}$
6. $\frac{-8}{13}+\frac{12}{13}i$
7. 13
8. $\frac{-254}{221}+\frac{667}{221}i$

Materials List/Setup

Station 1 none

Station 2 calculator

Station 3 none

Station 4 none

Discussion Guide

To support students in reflecting on the activities and to gather some formative information about student learning, use the following prompts to facilitate a class discussion to "debrief" the station activities.

Prompts/Questions

1. What is the square root of –1?
2. What is a complex number?
3. How do you find the conjugate of a complex number?
4. What is the product of a complex number $a + bi$ and its conjugate?

Think, Pair, Share

Have students jot down their own responses to questions, then discuss with a partner (who was not in their station group), and then discuss as a whole class.

Suggested Appropriate Responses

1. i
2. A complex number is one that can be expressed as $a + bi$, where a and b are real numbers.
3. If a complex number is expressed as $a + bi$, its conjugate is $a - bi$.
4. $a^2 + b^2$

Possible Misunderstandings/Mistakes

- Incorrectly multiplying polynomials
- Incorrectly finding the conjugate of a complex number
- Making simple arithmetical errors in adding, subtracting, multiplying, and dividing
- Not understanding the relationship between complex numbers in radical form and in $a + bi$ form
- Not recognizing that $i^4 = 1$ and $i^2 = -1$

NAME:

Number and Quantity

Set 1: Operations with Complex Numbers

Station 1

Race your partner to complete the addition and subtraction problems. Show all your work. When you have both finished, check each other's work.

1. $(1+3i)+(2+5i)$

2. $(3+7i)+(10-11i)$

3. $(18+3i)+(4+2i)$

4. $(16+2i)+(10+i)$

5. $(4i-7)+(12-4i)$

6. $(a+gi)+(2a+3gi)$

7. $(7i-8)-(18-2i)$

8. $(3i+2)-(3i-2)$

9. $(10-5i)-(3+2i)$

10. $(2+10i)-(6-7i)$

Station 2

Work with your partner to solve each problem. Show all your work. Use the calculator if necessary.

1. $(1+3i)(2+5i)$

2. $(3+7i)(4+2i)$

3. $(-1+2i)(3-2i)$

4. $\left(\frac{1}{4}+2i\right)(10+i)$

5. $(2i-3)4i$

6. $(3-i)(4+i)$

7. $(8+3i)(4-5i)$

8. $(10-2i^3)(4+1)$

9. $(9+2i)(9-2i)$

Station 3

Work with your group to identify the conjugate *c* of complex numbers and then solve each division problem. Show all your work.

1. $\dfrac{3+2i}{5-6i}$

2. $\dfrac{4+3i}{2+i}$

3. $\dfrac{5+2i}{3-2i}$

4. $\dfrac{3+2i}{3-2i}$

5. $\dfrac{7+3i}{7+3i}$

6. $\dfrac{8-3i}{2+i}$

7. $\dfrac{6-2i}{5+3i}$

8. $\dfrac{3-i}{4+2i}$

Station 4

Work with a group to solve each problem. State your answer in terms of $a + bi$. Show all your work.

1. $8+\sqrt{-\frac{1}{4}}$

2. $\sqrt{-16}+3$

3. $\sqrt{-9}-2$

4. $\sqrt{-25}\left(\sqrt{-16}-4\right)$

5. $\frac{4}{3+\sqrt{-9}}$

6. $\left(2\sqrt{-4}\right)\left(\frac{1}{3-\sqrt{-4}}\right)$

7. $\left(3+\sqrt{-4}\right)\left(3-\sqrt{-4}\right)$

8. $\frac{1}{4+\sqrt{-1}}+\frac{2+2\sqrt{-36}}{3-\sqrt{-4}}$

Seeing Structure in Expressions

Set 1: Factoring

Instruction

Goal: To provide opportunities for students to practice factoring quadratic equations

Common Core State Standard

A–SSE.3 Choose and produce an equivalent form of an expression to reveal and explain properties of the quantity represented by the expression.★

a. Factor a quadratic expression to reveal the zeros of the function it defines.

Student Activities Overview and Answer Key

Station 1

Given equations in the form $y = x^2 + bx + c$, $y = x^2 - bx + c$, $y = x^2 - bx - c$, or $y = x^2 + bx - c$, where b and c are integers, students work in groups to factor by grouping. Students will use algebra tiles as needed.

Answers

2. $(x + 4)(x + 4)$
3. $(x - 7)(x + 5)$
4. $(x - 1)(x + 9)$
5. $(x - 4)(x + 7)$
6. $(x + 2)(x - 1)$
7. $(x + 3)(x + 6)$
8. $(x - 4)(x + 11)$
9. $(x - 10)(x - 3)$
10. $(x - 5)(x + 5)$

Station 2

Given equations in the form $y = x^2 + bx + c$, $y = x^2 - bx + c$, $y = x^2 - bx - c$, or $y = x^2 + bx - c$, where b and c are real numbers, students work in pairs to factor by grouping.

Answers

1. $\left(x - \frac{1}{2}\right)\left(x - \frac{3}{8}\right)$
2. $\left(x + \frac{1}{5}\right)(x - 5)$

3. $(x-8)(x+3)$
4. $\left(x+\frac{1}{3}\right)\left(x+\frac{2}{7}\right)$
5. $\left(x-\frac{1}{6}\right)(x-2)$
6. $\left(x+\frac{4}{3}\right)(x-14)$
7. $(x+0.5)(x+0.3)$
8. $(x+0.2)(x-0.12)$

Station 3

Given equations in the form $y = ax^2 + bx + c$, $y = ax^2 - bx + c$, $y = ax^2 - bx - c$, or $y = ax^2 + bx - c$, where a, b, and c are real numbers, students work in groups to factor by grouping. Students will use algebra tiles as needed.

Answers

1. $4(x+2)(x-2)$
2. $3(x-1)(x-7)$
3. $5(2x+3)(x-6)$
4. $\frac{1}{2}(3x-1)(x-2)$
5. $\frac{2}{5}(x+4)(x-10)$
6. $7(x+3)(3x+1)$
7. $\frac{1}{9}(x-5)(x-2)$
8. $\frac{3}{4}(4x-7)(x+2)$

Station 4

Students are given a set of 12 index cards, each inscribed with one of the following expressions: 8, 6, $\frac{1}{2}$, 10, 3, $(x-5)$, $(x+2)$, $(x-3)$, $(x+7)$, $(x+1)$, $(3x+2)$, and $\left(x-\frac{1}{2}\right)$. They use grouping to factor a series of equations. Each card appears as a factor at least once in this series of equations. Then students combine their cards in pairs to come up with quadratic equations of their own.Z

Answers

1. $(x-5)(x+1)$
2. $8\left(x-\frac{1}{2}\right)(3x+2)$
3. $6(x-3)(3x+2)$
4. $\frac{1}{2}(x+7)(x+2)$
5. $10(x+1)\left(x-\frac{1}{2}\right)$
6. $3(x-5)(x-3)$
7. Answers will vary. Students should create three quadratic expressions that combine the factors on the cards.

Materials List/Setup

Station 1 algebra tiles

Station 2 none

Station 3 algebra tiles

Station 4 algebra tiles; 12 index cards with the following written on them (one expression per card): 8; 6; $\frac{1}{2}$; 10; 3; $(x-5)$; $(x+2)$; $(x-3)$; $(x+7)$; $(x+1)$; $(3x+2)$; $\left(x-\frac{1}{2}\right)$

Discussion Guide

To support students in reflecting on the activities and to gather some formative information about student learning, use the following prompts to facilitate a class discussion to "debrief" the station activities.

Prompts/Questions

1. What are factors?
2. What is distribution? How does it apply to binomial factors?
3. How do you factor a quadratic equation?
4. If you factor a quadratic equation that is set equal to 0, what points on the equation's graph do the factors represent? Why?

Think, Pair, Share

Have students jot down their own responses to questions, then discuss with a partner (who was not in their station group), and then discuss as a whole class.

Suggested Appropriate Responses

1. Factors are the quantities that are multiplied to produce a product.
2. Distribution is a property of real numbers that allows the multiplication of a term to a sum of terms. With a pair of binomials that are being multiplied together, the Distributive Property is used twice. Take the first term in the first binomial and multiply it by each term in the second binomial, adding the products. Then take the second term in the first binomial and multiply it by each term in the second binomial, adding all the products together.
3. Rewrite the equation in $y = ax^2 + bx + c$ form, with all x expressions and constants on the same side of the equation. Find the factors of a and the factors of c that combine to create b.
4. They represent the x-intercepts, because those are the points at which $y = 0$.

Possible Misunderstandings/Mistakes

- Incorrectly factoring quadratic expressions
- Incorrectly factoring constants and coefficients
- Not understanding factoring
- Not understanding polynomial factoring
- Not simplifying the equation before factoring
- Making simple arithmetical errors in factoring

NAME:

Seeing Structure in Expressions
Set 1: Factoring

Station 1

Work as a group to factor each equation. Use the algebra tiles if you wish. Show all your work.

1. $y = x^2 + 8x + 16$

2. $y = x^2 - 2x - 35$

3. $y = x^2 + 8x - 9$

4. $y = x^2 + 3x - 28$

5. $y = x^2 + x - 2$

6. $y = x^2 + 9x + 18$

7. $y = x^2 + 7x - 44$

8. $y = x^2 - 13x + 30$

9. $y = x^2 - 25$

Seeing Structure in Expressions
Set 1: Factoring

Station 2

Work in pairs to factor each equation. Show all your work. Check your work by using distribution to find the product of your factors.

1. $y = x^2 - \frac{7}{8}x + \frac{3}{16}$

2. $y = x^2 - \frac{24}{5}x - 1$

3. $y = x^2 - 5x - 24$

4. $y = x^2 + \frac{13}{21}x + \frac{2}{21}$

5. $y = x^2 - \frac{13}{6}x + \frac{1}{3}$

6. $y = x^2 - \frac{38}{3}x - \frac{56}{3}$

7. $y = x^2 + 0.8x + 0.15$

8. $y = x^2 + 0.08x - 0.024$

NAME:

Seeing Structure in Expressions
Set 1: Factoring

Station 3

Work as a group to factor each equation. Use the algebra tiles if you wish. Show all your work.

1. $y = 4x^2 - 16$

2. $y = 3x^2 - 24x + 21$

3. $y = 10x^2 - 45x - 90$

4. $y = \frac{3}{2}x^2 - \frac{7}{2}x + 1$

5. $y = \frac{2}{5}x^2 - \frac{12}{5}x - \frac{80}{5}$

6. $y = 21x^2 + 70x + 21$

7. $y = \frac{x^2}{9} - \frac{5}{9}x - \frac{2}{9}x + \frac{10}{9}$

8. $y = 3x^2 + \frac{3}{4}x - \frac{21}{2}$

Station 4

At this station, you will find algebra tiles and 12 index cards marked with the following expressions:

8	6	$\frac{1}{2}$	10	3	$(x - 5)$
$(x + 2)$	$(x - 3)$	$(x + 7)$	$(x + 1)$	$(3x + 2)$	$\left(x - \frac{1}{2}\right)$

Work as a group to factor each equation below, using the index cards provided. You will also use the factors later in the activity. Use the algebra tiles if you wish. Show all your work.

1. $y = x^2 - 4x - 5$

2. $y = 24x^2 + 4x - 8$

3. $y = 18x^2 - 42x - 36$

4. $y = \frac{1}{2}x^2 + \frac{9}{2}x + 7$

5. $y = 10x^2 + 5x - 5$

6. $y = 3x^2 - 24x + 45$

7. Combine your factor cards to form three different quadratic equations. Write your equations below.

Seeing Structure in Expressions

Set 2: Quadratic Transformations in Vertex Form

Instruction

Goal: To provide opportunities for students to analyze the relationship between the equation of a parabola and its graph

Common Core State Standards

A–SSE.3 Choose and produce an equivalent form of an expression to reveal and explain properties of the quantity represented by the expression.★

b. Complete the square in a quadratic expression to reveal the maximum or minimum value of the function it defines.

F–IF.7 Graph functions expressed symbolically and show key features of the graph, by hand in simple cases and using technology for more complicated cases.★

a. Graph linear and quadratic functions and show intercepts, maxima, and minima.

F–IF.8 Write a function defined by an expression in different but equivalent forms to reveal and explain different properties of the function.

a. Use the process of factoring and completing the square in a quadratic function to show zeros, extreme values, and symmetry of the graph, and interpret these in terms of a context.

F–BF.3 Identify the effect on the graph of replacing $f(x)$ by $f(x) + k$, $k\,f(x)$, $f(kx)$, and $f(x + k)$ for specific values of k (both positive and negative); find the value of k given the graphs. Experiment with cases and illustrate an explanation of the effects on the graph using technology. *Include recognizing even and odd functions from their graphs and algebraic expressions for them.*

Student Activities Overview and Answer Key

Station 1

Given equations in the form $y = x^2 + k$ and $y = (x - h)^2$, where h and k are integers, students graph a series of parabolas, finding the y-intercept and the axis of symmetry. They explore the relationship between the value of h and k and the position of the parabola with respect to the x- and y-axes. Students should also begin to understand the relationship between the equation of the parabola and the axis of symmetry.

Answers

2.

x	y
0	0
1	1
2	4
3	9
–1	1
–2	4
–3	9

3.

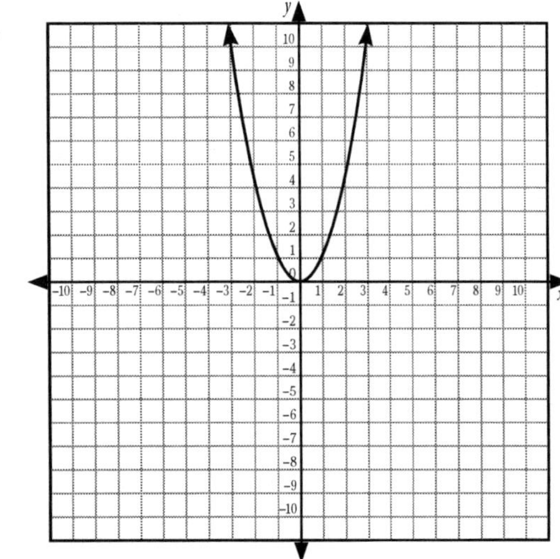

4. (0, 0)

5. (0, 2)

6. $y = x^2 + 2$

7. $y = x^2 - 5$

8.

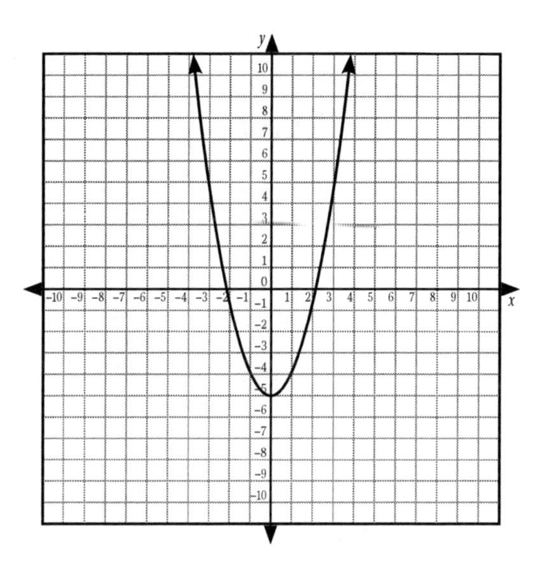

9.

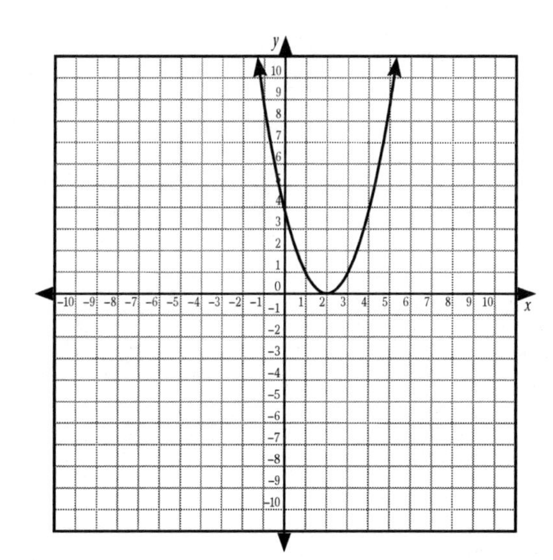

10. $x = 2$

11. $x = -3$

Station 2

Given equations in the form $y = ax^2$, students graph parabolas. Students compare graphs to explore the relationship between the coefficient of x and the width of the parabola.

Answers

1.

x	y
0	0
1	3
2	12
3	27
–1	3
–2	12
–3	27

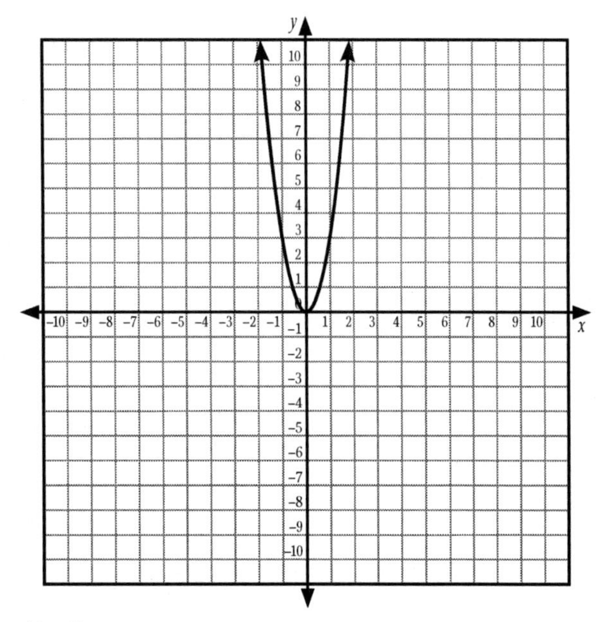

2. (0, 0)
3. $x = 0$

4.

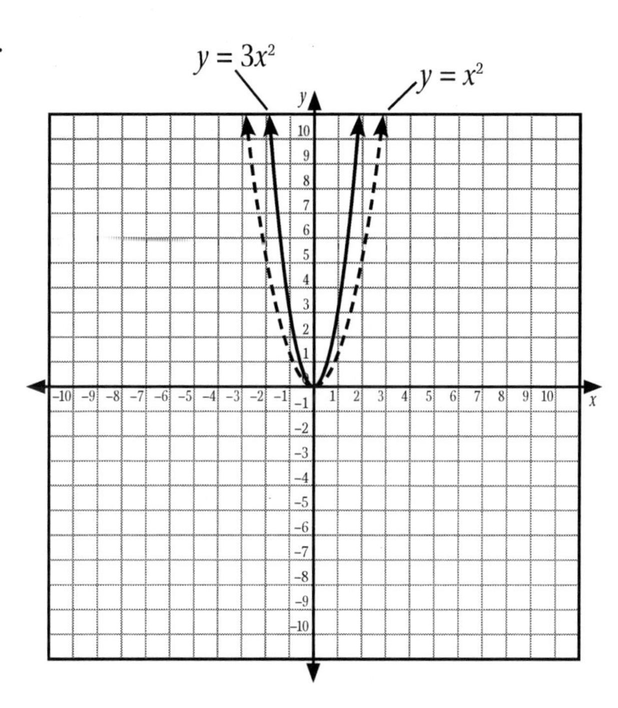

The parabola $3x^2$ is narrower than the parabola x^2.

5–6.

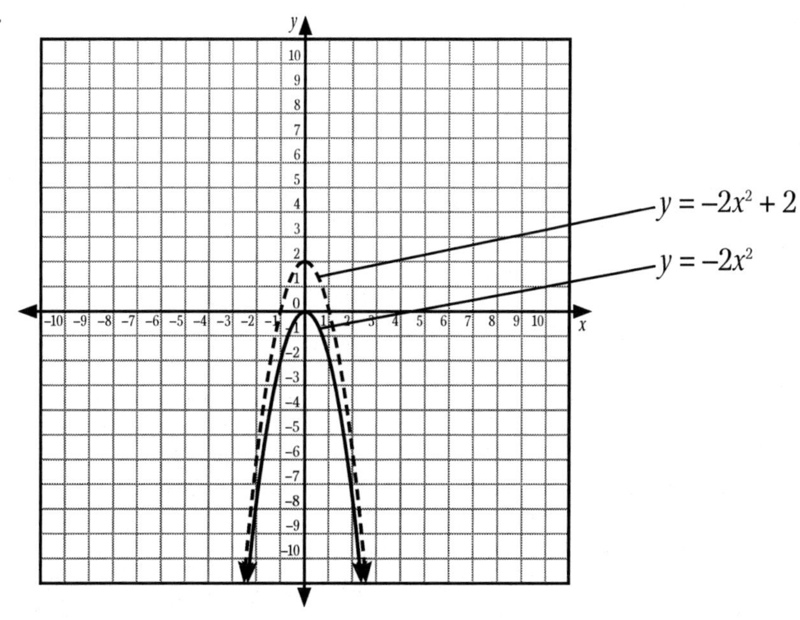

7. (0, 2)
8. The parabola moves vertically.
9. The parabola changes in width.

Station 3

Given equations in the form $y = (x - h)^2 + k$, students graph parabolas. Students find the y-intercept and the axis of symmetry from both the graph and the equation, and begin working towards an understanding of the vertex of a parabola.

Answers

1.

x	y
0	5
1	3
2	5
3	11
4	21
–1	11
–2	21
–3	35

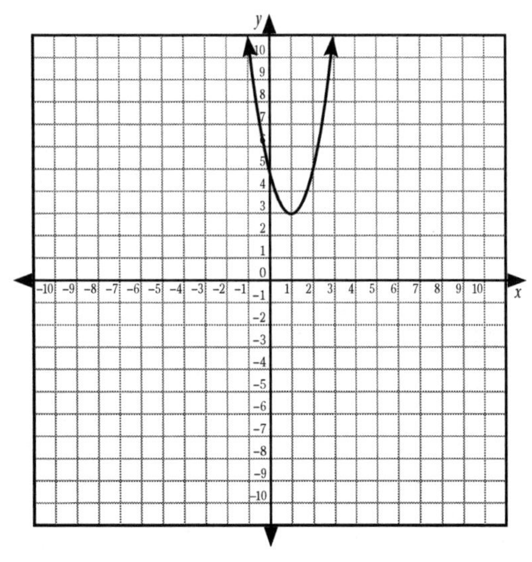

2. $x = 1$

3. (0, 5)

4. The parabola would open downward.

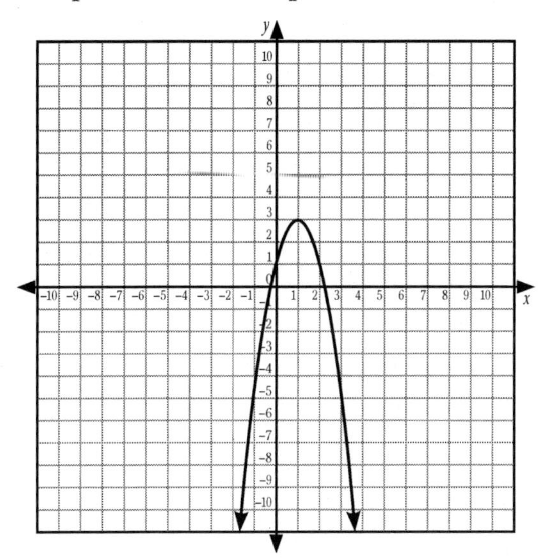

5. (0, 3)

$$y = \frac{1}{2}(x-2)^2 + 1$$

$$y = \frac{1}{2}(0-2)^2 + 1$$

$$y = \frac{1}{2}(4) + 1$$

$$y = 3$$

6. It will be wider, because the higher the coefficient of the x^2 expression, the narrower the parabola. The coefficient of the second x^2 expression is 1, which is higher than ½, the coefficient of the first x^2 expression.

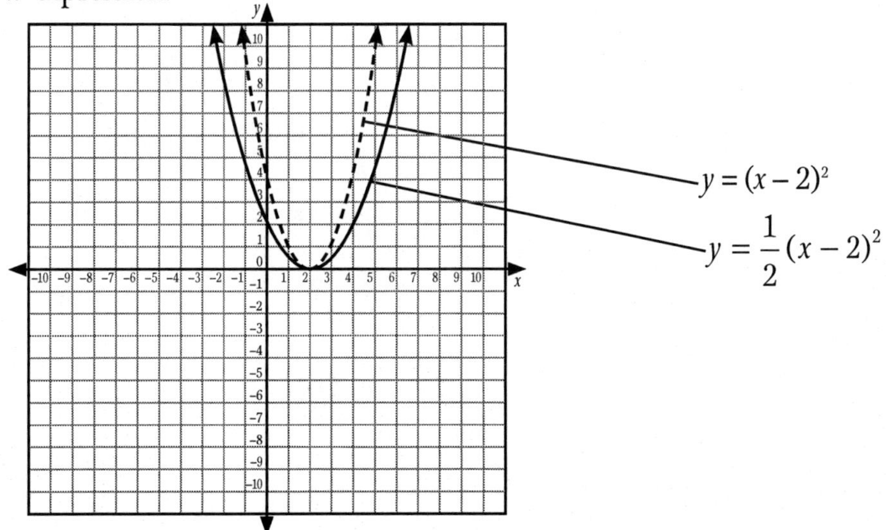

7. (0, 17)

8. $y = 3(x - 2)^2 + 10$

9. The new y-intercept is (0, 22), so solve for the value of a.

$$y = 3(x - 2)^2 + a$$
$$22 = 3(0 - 2)^2 + a$$
$$22 = 3(4) + a$$
$$22 = 12 + a$$
$$10 = a$$

Station 4

Students use two methods (completing the square and finding the midpoint of the x-intercepts) to convert the equations of parabolas from quadratic form to vertex form. They graph to check their work and to understand the correlation between the different forms and the graph. Students should recognize that a parabola's axis of symmetry always runs through its vertex. They should also understand the relationship between the coordinates of the vertex and the equation in vertex form.

Answers

1.

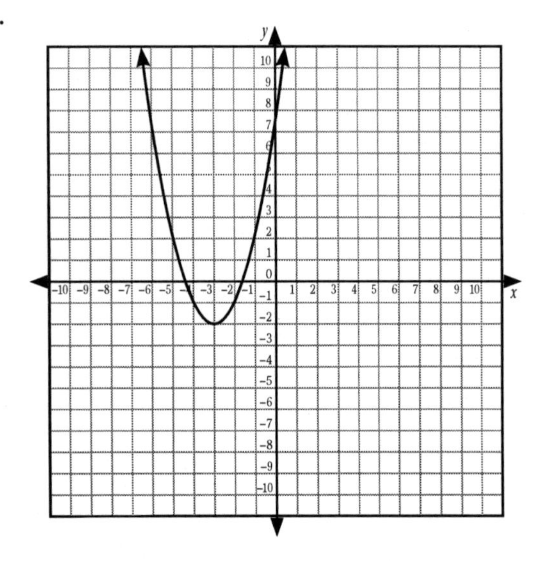

2. $x = -3$

3. $y = x^2 + 6x + 7$

$$y = x^2 + 6x + 7 + 2 - 2$$
$$y = (x^2 + 6x + 9) - 2$$
$$y = (x + 3)^2 - 2$$

4. (–3, –2)
5. (4, 1)
6. $y = \frac{x^2}{2} - 4x + 9$

   $y = \frac{x^2}{2} - 4x + 8 + 1$

   $y = \frac{1}{2}(x^2 - 8x + 16) + 1$

   $y = \frac{1}{2}(x - 4)^2 + 1$
7. (0, 9)
8. $x = 4$
9. Yes. The parabola opens out from the vertex. The vertex contains the only y-coordinate that is not repeated in the range.
10. Because the parabola is symmetrical, the axis of symmetry will intersect the midpoint of the line between the x-intercepts. The midpoint is at (0, 0). That means the x-coordinate at the vertex must be 0, because the axis of symmetry intersects the vertex. If $x = 0$, $y = -4$, so the coordinates of the vertex are (0, –4).

Materials List/Setup

Station 1 graph paper

Station 2 colored pencils or pens; graph paper

Station 3 colored pencils or pens; graph paper

Station 4 graph paper

Discussion Guide

To support students in reflecting on the activities and to gather some formative information about student learning, use the following prompts to facilitate a class discussion to "debrief" the station activities.

Prompts/Questions

1. What is a function's axis of symmetry? Does every parabola have one?
2. What is a *y*-intercept?
3. How do you find the coordinates of a parabola's *y*-intercept?
4. Compare the equations $y = a(x - h)^2 + k$ and $y = ax^2 - 2axh + ah^2 + k$. Do you think they express the same thing? How could you find out?

Think, Pair, Share

Have students jot down their own responses to questions, then discuss with a partner (who was not in their station group), and then discuss as a whole class.

Suggested Appropriate Responses

1. An axis of symmetry is the line that divides the graph of the function into two symmetrical halves. Every parabola has an axis of symmetry.
2. A *y*-intercept is the point at which a function crosses the *y*-axis.
3. Set *x* equal to 0 and solve the function for *y*.
4. Students should multiply out the equation in vertex form to find the equation in quadratic form.

Possible Misunderstandings/Mistakes

- Incorrectly calculating the value of *y*-coordinates from *x*-coordinates
- Incorrectly graphing parabolas, either from incorrect calculations or from a misunderstanding of graphing itself
- Not understanding the definition of the vertex
- Assuming that the vertex is unrelated to the axis of symmetry
- Incorrectly factoring quadratic equations
- Making simple arithmetical errors in completing the square

- Not understanding the arithmetical manipulations involved in completing the square
- Confusing the y-intercept with the x-intercept
- Confusing the vertex coordinates h and k
- English language learners may struggle with the questions that ask for written explanations. Encourage these students to write out the numeric operations involved and then describe their work out loud.

NAME:

Seeing Structure in Expressions

Set 2: Quadratic Transformations in Vertex Form

Station 1

Work as a group to answer the questions. Construct graphs without the aid of a graphing calculator. Show all your work and label the axes of each graph.

1. Given the parabola $y = x^2$, complete the table below with the y coordinates for the following values of x.

x	y
0	
1	
2	
3	
–1	
–2	
–3	

2. Use the coordinates from your table to graph the parabola on graph paper.

3. What are the coordinates for the parabola's y-intercept?

4. Look at the parabola below. What is its y-intercept?

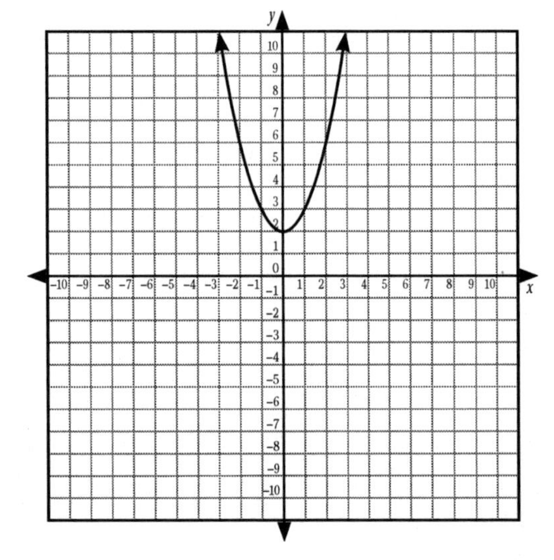

continued

5. What is the equation for this parabola?

6. How would you write the equation for a similar parabola of y-intercept $(0, -5)$?

7. Graph the parabola from problem 6.

8. Graph the parabola $y = (x - 2)^2$.

9. What is the equation for the axis of symmetry of this parabola?

10. Without graphing, predict the equation for the axis of symmetry of the parabola $y = (x + 3)^2$.

Station 2

Work with your group to explore the relationship between a quadratic function and its graph.

1. Given the equation $y = 3x^2$, complete the table with the values of y and graph the parabola.

x	y
0	
1	
2	
3	
–1	
–2	
–3	

2. What are the coordinates of this parabola's y-intercept?

3. What is the equation of its axis of symmetry?

4. On the graph from problem 1, draw the parabola $y = x^2$ in a contrasting color. In words, compare the two parabolas.

5. Graph the parabola $y = -2x^2$. Complete the table if you need a reference.

x	y
0	
1	
2	
3	
–1	
–2	
–3	

continued

6. On the same graph, in a contrasting color, graph the parabola $y = -2x^2 + 2$. Label each parabola.

7. What are the coordinates of the y-intercept of $y = -2x^2 + 2$?

8. What happens to the graph of a parabola when you add a numeric constant to its equation, as in problem 6?

9. What happens to the graph of a parabola when the x^2 expression is given a numeric coefficient, as in problems 1 and 5? (*Hint*: Compare the parabola $y = 3x^2$ to $y = x^2$.)

Station 3

Work with your group to answer the following questions.

1. Complete the table for the parabola $y = 2(x - 1)^2 + 3$. Graph the parabola on graph paper.

x	y
0	
1	
2	
3	
4	
–1	
–2	
–3	

2. What is the equation for this parabola's axis of symmetry?

3. What are the coordinates of this parabola's y-intercept?

4. How would this graph change if the parabola's equation changed to $y = -2(x - 1)^2 + 3$? Graph the new parabola to check your answer.

5. What are the coordinates of the y-intercept of the parabola $y = \frac{1}{2}(x-2)^2 + 1$?

continued

6. Do you think that the graph of $y = \frac{1}{2}(x-2)^2$ will be wider or narrower than the graph of $y = (x-2)^2$? Why? Graph both parabolas, in contrasting colors, to check your answer.

7. Look at the graph below. The equation for this parabola is $y = 3(x-2)^2 + 5$. What is its y-intercept?

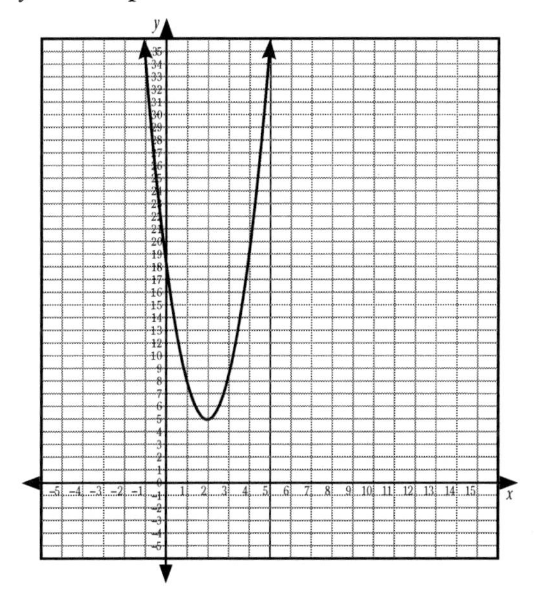

8. How would you write the equation for a similar parabola with a y-intercept 5 units higher? Show your work. Write out an explanation in words if necessary.

Seeing Structure in Expressions

Set 2: Quadratic Transformations in Vertex Form

Station 4

Work with your group to answer the following questions.

1. Graph the parabola $y = x^2 + 6x + 7$ on graph paper.

2. Give the equation for its axis of symmetry.

3. *Optional*: Complete the square to give the equation for the parabola in vertex form. Show your work.

4. What are the coordinates of the vertex of this parabola?

5. Look at the graph below. What are the coordinates of the vertex of this parabola?

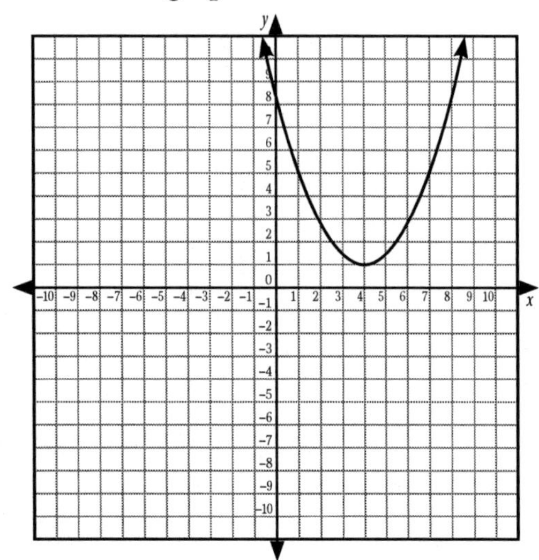

continued

NAME:

Seeing Structure in Expressions
Set 2: Quadratic Transformations in Vertex Form

6. The equation for this parabola is $y = \frac{x^2}{2} - 4x + 9$. Find the vertex in order to convert the equation to vertex form. Show your work.

7. What are the coordinates of the *y*-intercept?

8. What is the equation for the axis of symmetry?

9. Does a parabola's axis of symmetry always run through its vertex? Why or why not?

10. Look at the graph below, which shows the parabola $y = x^2 - 4$. The coordinates of the parabola's *x*-intercepts are (2, 0) and (–2, 0). How could you use this information to find the coordinates of the parabola's vertex? Explain, showing your work.

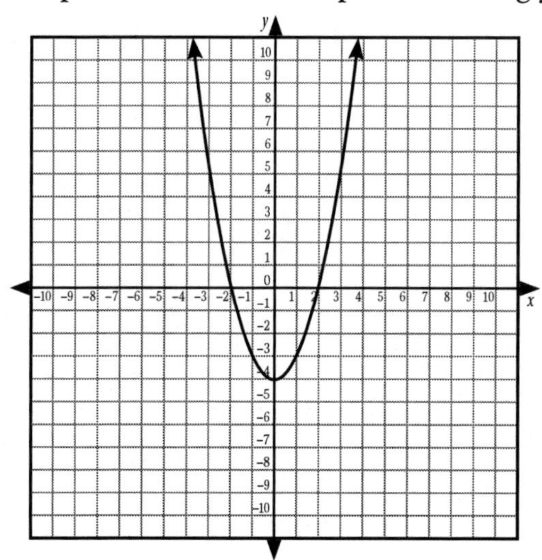

Reasoning with Equations and Inequalities

Set 1: Solving Quadratics

Instruction

Goal: To provide opportunities for students to solve quadratic equations with real and complex solutions

Common Core State Standards

N–CN.7 Solve quadratic equations with real coefficients that have complex solutions.

A–SSE.3 Choose and produce an equivalent form of an expression to reveal and explain properties of the quantity represented by the expression.★

a. Factor a quadratic expression to reveal the zeros of the function it defines.

A–REI.4 Solve quadratic equations in one variable.

b. Solve quadratic equations by inspection (e.g., for $x^2 = 49$), taking square roots, completing the square, the quadratic formula and factoring, as appropriate to the initial form of the equation. Recognize when the quadratic formula gives complex solutions and write them as $a \pm bi$ for real numbers a and b.

Student Activities Overview and Answer Key

Station 1

Students work with a partner to solve the quadratic equations by factoring. Students may use algebra tiles as needed.

Answers

3. $x = -\frac{1}{2}, x = 3$
4. $x = \frac{1}{3}, x = -2$
5. $x = 7, x = 3$
6. $x = -\frac{4}{5}, x = 1$
7. $x = -3, x = 3$
8. $x = 4, x = 2$
9. $x = 2$

Station 2

Students work alone or in groups to solve the quadratic equations by factoring. They should be increasingly comfortable finding real roots independently.

Answers

1. $x = -8, x = \frac{1}{9}$
2. $x = 1, x = \frac{1}{12}$
3. $x = -2, x = 6$
4. $x = 1, x = -\frac{7}{5}$
5. $x = 9$
6. $x = 2, x = 12$
7. $x = -\frac{2}{7}, x = \frac{5}{2}$

Station 3

Students work in groups to solve quadratic equations using the quadratic formula. They will use a graphing calculator. They begin working with irrational roots.

Answers

1. $x = \frac{2 \pm 13.266i}{3}$
2. $x = 3.618, x = 1.382$
3. $x = \frac{3 \pm 22.428i}{32}$
4. $x = \frac{1 \pm 3.872i}{4}$
5. $x = -9.3589$, $x = -0.641$

Station 4

Students work in pairs to solve quadratic equations using factoring and the quadratic formula. They will use a graphing calculator. Students may begin to recognize the relationship between the discriminant and the type of roots.

Answers

1. $x = \dfrac{5 \pm 3.873i}{20}$
2. $x = \dfrac{6 \pm 10.392i}{2}$
3. $x = 8$
4. $x = -1.146, x = -7.854$
5. $x = -5, x = -2$
6. $x = -1, x = -2$
7. Students may not have noticed any patterns. Some students may notice the relationship between the discriminant and the type of roots the equation has.

Materials List/Setup

Station 1 algebra tiles

Station 2 none

Station 3 graphing calculator

Station 4 graphing calculator

Discussion Guide

To support students in reflecting on the activities and to gather some formative information about student learning, use the following prompts to facilitate a class discussion to "debrief" the station activities.

Prompts/Questions

1. What is the quadratic formula?
2. What is a real number?
3. What is a complex number?
4. What is the value of i?
5. What is the relationship among factors, x-intercepts, roots, and solutions of a quadratic?

Think, Pair, Share

Have students jot down their own responses to questions, then discuss with a partner (who was not in their station group), and then discuss as a whole class.

Suggested Appropriate Responses

1. If $y = ax^2 + bx + c$, then $x = \frac{-b \pm \sqrt{b^2 - 4ac}}{2a}$.
2. A real number is any number that can be expressed as a decimal.
3. A complex number is any number expressed as the sum or difference of a real number and an imaginary number, or $a + bi$.
4. $\sqrt{-1}$
5. A quadratic equation written in factored form shows the factors of the quadratic equation. By taking these factors and setting each of them equal to 0 and solving for x, you are finding the roots or the solutions of the quadratic equation. The roots of the equation are the x-intercepts, or where the graph crosses the x-axis. These are also said to be the solutions. If the equation is in the form "$y =$" or "$f(x) =$", you will be setting that side of the equation equal to 0. This is the equivalent of finding the x-intercepts. Use the Zero Product Property to set each factor equal to 0 and solve for the variable.

Possible Misunderstandings/Mistakes

- Incorrectly factoring quadratic expressions
- Incorrectly factoring constants and coefficients
- Not understanding factoring
- Not understanding polynomial factoring
- Making simple arithmetical errors in factoring or in applying the quadratic formula
- Incorrectly applying the quadratic formula
- Rounding inaccurately
- Not setting $ax^2 + bx + c$ equal to 0

Reasoning with Equations and Inequalities

Set 1: Solving Quadratics

Station 1

Work with a partner to solve the quadratic equations by factoring. Use algebra tiles as needed. Show all your work.

1. $2x^2 = 5x + 3$

2. $3x^2 + 5x = 2$

3. $x^2 - 10x = -21$

4. $5x^2 = x + 4$

5. $x^2 - 9 = 0$

6. $2x^2 = 2(6x - 8)$

7. $7x^2 = 28(x - 1)$

Station 2

Work alone or with your group to solve the quadratic equations by factoring. Show all your work.

1. $-9x^2 + 8 = 71x$

2. $4x^2 = 13x - 1$

3. $\frac{x^2}{4} - x = 3$

4. $5x^2 = 7 - 2x$

5. $x^2 = -81 + 18x$

6. $2x - x^2 = 24 - 12x$

7. $31x = 14x^2 - 10$

NAME:

Reasoning with Equations and Inequalities

Set 1: Solving Quadratics

Station 3

Work with a partner to solve the problem using the quadratic formula. Use the graphing calculator to estimate square roots to the thousandths. State irrational roots in terms of *i.*

1. $y = 3x^2 - 4x + 16$

2. $y = x^2 - 5x + 5$

3. $y = 8x^2 - \frac{3}{2}x + 4$

4. $y = 2x^2 - x + 2$

5. $y = \frac{x^2}{2} + 5x + 3$

Station 4

Work alone or with your group to solve using either factoring or the quadratic formula. Use the graphing calculator to estimate square roots to the thousandths. State irrational roots in terms of *i*.

1. $y = 10x^2 - 5x + 1$

2. $y = \frac{x^2}{3} - 2x + 12$

3. $y = x^2 - 16x + 64$

4. $y = x^2 + 9x + 9$

5. $y = x^2 + 7x + 10$

6. $y = 3x^2 + 9x + 6$

7. Do you notice any patterns about the value of $\sqrt{b^2 - 4ac}$ and the roots of the equation? If so, what are they?

Arithmetic with Polynomials and Rational Expressions

Set 1: Operations with Polynomials

Instruction

Goal: To provide opportunities for students to develop concepts and skills related to adding, subtracting, multiplying, and dividing polynomials

Common Core State Standard

A–APR.1 Understand that polynomials form a system analogous to the integers, namely, they are closed under the operations of addition, subtraction, and multiplication; add, subtract, and multiply polynomials.

Student Activities Overview and Answer Key

Station 1

Students will be given 20 blue algebra tiles, 20 red algebra tiles, 20 green algebra tiles, and 20 yellow algebra tiles. Students work together to model polynomials with algebra tiles. Then they add the polynomials using the algebra tiles.

Answers

1. $8x^2 + xy + 5y^2$
2. Answers will vary. Possible answer: We combined same-color algebra tiles.
3. Answers will vary. Possible answer: We used the Zero Product Property to find pairs of the same colored algebra tiles that canceled each other out.
4. Zero Product Property
5. $12y^2 - 12xy - 5x^2 - 4$
6. Answers will vary.
7. Answers will vary. Possible answer: We used the Zero Product Property.
8. $5a^3 - 3a^2b^2 + 10b^3$
9. $10x^2 - y^2 - 15xy + 4$
10. $16c^3 - 8a^3 + 3ac^2 - 7$

Station 2

Students will be given 20 blue algebra tiles, 20 red algebra tiles, 20 green algebra tiles, and 20 yellow algebra tiles. Students will work together to model polynomials with algebra tiles. Then they will subtract polynomials using the algebra tiles.

Answers

1. $5x^2 + 5xy + 4y^2$
2. Answers will vary.
3. $3x^2$, $2xy$, $2y^2$
4. $-5x^2 - 5xy - 4y^2$
5. No, because the sign of the terms in the second polynomial changes to the opposite sign.
6. $7x^2 + 5xy + 11y^2$
7. Answers will vary. Possible answer: We matched like terms and then performed subtraction.
8. Answers will vary. Possible answer: We added negative terms because subtracting a negative number is the same as adding a positive of that number.
9. $-2a^4 - 4a^2b^2 + 6b^3 + 6$
10. $6c^2 - 6bc - 18$

Station 3

Students will be given a number cube. Students will use the number cube to populate coefficients of polynomials. Then they will multiply polynomials using the Distributive Property.

Answers

1. Answers will vary. Possible answer: $2x$ and $(3x + y - 2)$
2. Distributive Property
3. Answers will vary. Possible answer: $\begin{array}{c} 2x(3x + y - 2) \\ 6x^2 + 2xy - 4x \end{array}$
4. Answers will vary. Possible answer: $-3x^2$ and $(-4x + 7xy - 8)$
5. Answers will vary. Possible answer: $12x^3 - 21x^3y + 24x^2$
6. It changed to the opposite sign because we multiplied each term by -1.
7. $(x + 3)$ and $(x - 4)$
8. Distributive Property
9. $(x + 3)(x - 4)$

   $x^2 - 4x + 3x - 12$

   $x^2 - x - 12$
10. We combined $-4x$ and $3x$.

Station 4

Students will be given six index cards with the following polynomials written on them:

$3xy^2$	$18x^2 - 7x + 4$	$33xy^5 - 3x^2y^2 - 21xy^2$
$2x$	$-3y^2$	$-24y^5 + 6y^3 - 12$

Students will work together to match polynomials and monomials that when divided by each other yield a specific quotient. Then students will perform synthetic division to divide a polynomial by a binomial.

Answers

1. $\dfrac{18x^2 - 7x + 4}{2x} = 9x - \dfrac{7}{2} + \dfrac{2}{x}$
2. $\dfrac{-24y^5 + 6y^3 - 12}{-3y^2} = 8y^3 - 2y + \dfrac{4}{y^2}$
3. $\dfrac{33xy^5 - 3x^2y^2 - 21xy^2}{3xy^2} = 11y^3 - x - 7$
4. Answers will vary. Possible answer: We divided each term by the monomial using the quotient rule for exponents.
5. No, because you have to divide by a binomial instead of a monomial.
6. The degree of the polynomial is 3. The quotient will have 3 terms.
7. Find the solution of the binomial, which is 1. Write this in the left hand box. Write the coefficients of the variables in a row. Bring down the first coefficient. Multiply this coefficient by 1. Add this product to the second coefficient. Repeat this process through all the coefficients. The last number is the remainder.
8. $2x^3 + 6x^2 + 7x + \dfrac{4}{x - 1}$
9. $4x^3 + 5x^2 + 15x + 31 + \dfrac{57}{x - 2}$

Materials List/Setup

Station 1 20 blue algebra tiles; 20 red algebra tiles; 20 green algebra tiles; 20 yellow algebra tiles

Station 2 20 blue algebra tiles; 20 red algebra tiles; 20 green algebra tiles; 20 yellow algebra tiles

Station 3 number cube

Station 4 six index cards with the following polynomials written on them:
$3xy^2$; $18x^2 - 7x + 4$; $33xy^5 - 3x^2y^2 - 21xy^2$; $2x$; $-3y^2$; $-24y^5 + 6y^3 - 12$

Discussion Guide

To support students in reflecting on the activities and to gather some formative information about student learning, use the following prompts to facilitate a class discussion to "debrief" the station activities.

Prompts/Questions

1. How do you add polynomials?
2. How do you subtract polynomials?
3. What happens to the exponents of the variables when you add or subtract polynomials?
4. How do you multiply polynomials?
5. How do you deal with the exponents of the variables when multiplying polynomials?
6. How do you divide a polynomial by a monomial?
7. How do you divide a polynomial by a binomial?
8. How do you deal with the exponents of the variables when dividing polynomials?

Think, Pair, Share

Have students jot down their own responses to questions, then discuss with a partner (who was not in their station group), and then discuss as a whole class.

Suggested Appropriate Responses

1. Add like terms.
2. Subtract like terms of the second polynomial from like terms in the first polynomial.
3. Exponents remain the same in addition and subtraction of polynomials.
4. Multiply each term in the first polynomial by each term in the second polynomial using the Distributive Property.
5. Use the product rule on the exponents.
6. Divide each term in the polynomial by the monomial.
7. Use synthetic division.
8. Use the quotient rule on the exponents.

Possible Misunderstandings/Mistakes

- Incorrectly adding exponents when adding polynomials
- Incorrectly subtracting exponents when subtracting polynomials
- Not using the product rule on exponents when multiplying polynomials
- Not using the quotient rule on exponents when dividing polynomials
- Not using synthetic division when dividing by a binomial
- Not realizing that the last number in synthetic division is the remainder

Station 1

At this station, you will find 20 blue algebra tiles, 20 red algebra tiles, 20 green algebra tiles, and 20 yellow algebra tiles. Work as a group to model each polynomial by placing the tiles next to the polynomials. Then find the sum. Write your answer in the space provided below each problem.

- Use the blue algebra tiles to model the x^2 term.
- Use the red algebra tiles to represent the xy term.
- Use the green algebra tiles to represent the y^2 term.
- Use the yellow algebra tiles to represent the constant.

1. Given: $\begin{array}{r} 3x^2 + 2xy + 2y^2 \\ + \quad 5x^2 - xy + 3y^2 \\ \hline \end{array}$. Model the polynomial and find the sum.

2. How did you use the algebra tiles to model the problem?

3. How did you model the $-xy$ term?

4. What property did you use on the xy terms?

5. Model the following problem using the algebra tiles. Show your work, and write your answer in the space below.

   $(4y^2 - 12xy + 5x^2) + (-10x^2 + 8y^2 - 4)$

continued

6. How did you use the algebra tiles to model problem 5?

7. How did you deal with negative terms during addition?

Work together to add each polynomial. Show your work, and write your answer in the space below each problem.

8. Given:

$$\begin{array}{r} 2a^3 + a^2b^2 + 3b^3 \\ +\quad 3a^3 - 4a^2b^2 + 7b^3 \\ \hline \end{array}$$

9. $-10xy - 3 + 2x^2 - 5y^2 + 4y^2 + 8x^2 - 5xy + 7$

10. $8c^3 + 3ac^2 + 4a^3 + 8c^3 - 12a^3 - 7$

Station 2

At this station, you will find 20 blue algebra tiles, 20 red algebra tiles, 20 green algebra tiles, and 20 yellow algebra tiles. Work as a group to model each polynomial by placing the tiles next to the polynomials. Then find the difference. Write your answer in the space provided below each problem.

- Use the blue algebra tiles to model the x^2 term.
- Use the red algebra tiles to represent the xy term.
- Use the green algebra tiles to represent the y^2 term.
- Use the yellow algebra tiles to represent the constant.

1. Given: $\begin{array}{r} 8x^2 + 7xy + 6y^2 \\ -(3x^2 + 2xy + 2y^2) \\ \hline \end{array}$. Model the polynomial and find the difference.

2. How did you use the algebra tiles to model the problem?

3. What terms in the bottom polynomial does the subtraction sign apply to?

4. Find the difference: $\begin{array}{r} 3x^2 + 2xy + 2y^2 \\ -\ (8x^2 + 7xy + 6y^2) \\ \hline \end{array}$. Write your answer in the space below.

continued

5. Is your answer from problem 1 the same as your answer from problem 4? Why or why not?

6. Model the subtraction problem below using the algebra tiles, then solve. Show your work, and write your answer in the space below.

$$\begin{array}{r} 2x^2 + 5y^2 + 9xy \\ -\ (4xy - 5x^2 - 6y^2) \\ \hline \end{array}$$

7. How did you arrange the algebra tiles to model problem 6?

8. How did you deal with negative terms during subtraction?

9. Work together to subtract each polynomial. Show your work, and write your answer in the space below each problem.

$$\begin{array}{r} a^4 - a^2b^2 + 4b^3 + 8 \\ -\ (3a^4 + 3a^2b^2 - 2b^3 + 2) \\ \hline \end{array}$$

10. Subtract $8c^2 + 2bc + 10$ from $-4bc + 14c^2 - 8$.

Station 3

At this station, you will find a number cube. As a group, roll the number cube. Write the result in the box below.

Given: $\square x(3x + y - 2)$

1. Identify the two polynomials above.

2. What property can you use to multiply these polynomials?

3. Multiply the polynomials. Show your work.

As a group, roll the number cube. Write the result in the box below.

Given: $-\square x^2(-4x + 7xy - 8)$

4. Identify the two polynomials above.

5. Multiply the polynomials. Show your work.

continued

6. What happened to the signs of each term of the polynomial in the parentheses? Explain your answer.

Given: $(x + 3)(x - 4)$

7. Identify the two polynomials above.

8. What method can you use to multiply these polynomials?

9. Multiply the polynomials. Show your work.

10. What extra steps did you take when multiplying $(x + 3)(x - 4)$ versus $-\square x^2(-4x + 7xy - 8)$?

Station 4

At this station, you will find six index cards with the following polynomials written on them:

$3xy^2$	$18x^2 - 7x + 4$	$33xy^5 - 3x^2y^2 - 21xy^2$
$2x$	$-3y^2$	$-24y^5 + 6y^3 - 12$

Shuffle the cards. Work as a group to match the polynomials that when divided yield each quotient below. (*Hint*: Place the monomials in the denominator.)

1. $9x - \frac{7}{2} + \frac{2}{x}$

2. $8y^3 - 2y + \frac{4}{y^2}$

3. $11y^3 - x - 7$

4. What strategy did you use for problems 1–3?

Given: $(2x^3 + 4x^2 + x - 3) \div (x - 1)$

5. Can you use the same strategy to divide the polynomials above as you did in problems 1–3? Why or why not?

6. What is the degree of $(2x^3 + 4x^2 + x - 3)$? What does this degree tell you about how many terms the quotient will have?

continued

Follow these steps to use synthetic division to find the quotient of $(2x^3 + 4x^2 + x - 3) \div (x - 1)$.

Step 1: Set the binomial equal to zero and solve for x. Show your work.

Step 2: Use your answer from Step 1 and write it in the first box on the left in the illustration below.

☐	☐	☐	☐	☐
	☐	2	6	7
	2	6	7	4

Step 3: Write the coefficients of each term in order of left to right in the top row of boxes in the illustration under Step 2.

Step 4: The first coefficient (in this case, 2 of $2 \bullet 1$) is always written in the box underneath it.

Step 5: The boxes that are already filled in show synthetic division.

7. Derive the process of synthetic division based on the example above.

Step 6: The last number in the bottom row, 4, is known as the remainder and can be written as $\frac{4}{x-1}$.

8. Based on steps 1–6, what is the answer to $(2x^3 + 4x^2 + x - 3) \div (x - 1)$ using synthetic division?

9. Use synthetic division to find $(4x^4 - 3x^3 + 5x^2 + x - 5) \div (x - 2)$. Show your work.

Functions

Set 1: Graphing Quadratic Equations

Instruction

Goal: To provide opportunities for students to develop concepts and skills related to graphing quadratic equations and functions

Common Core State Standard

F–IF.7 Graph functions expressed symbolically and show key features of the graph, by hand in simple cases and using technology for more complicated cases.★

a. Graph linear and quadratic functions and show intercepts, maxima, and minima.

Student Activities Overview and Answer Key

Station 1

Students will be given graph paper and a ruler. Students will derive how to find the vertex of the graph of a quadratic function. Then they will find the x-intercept and y-intercept of the function. They will graph the quadratic function and describe its shape.

Answers

2. $f(x) = x^2 + 6x + 9$
3. $a = 1$; $b = 6$; and $c = 9$
4. $\frac{-b}{2a}$
5. $f\left(\frac{-b}{2a}\right)$
6. $\left(\frac{-b}{2a}, f\left(\frac{-b}{2a}\right)\right)$; (–3, 0)
7. Use the vertex as the axis of symmetry and a table of values to determine the corresponding y-value when $x = 0$.
8. Set $x = 0$ and solve for y.
9. (–3, 0) and (0, 9)

10.

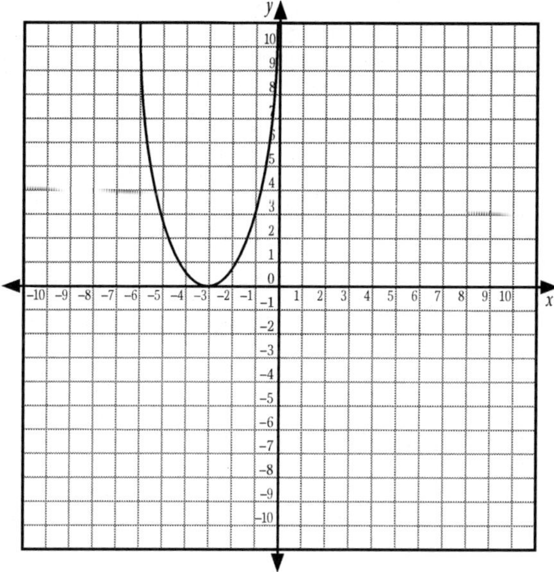

11. Parabola; the square term gives two x-values for each y-value due to the nature of square roots.

Station 2

Students will be given a graphing calculator. Students will use the graphing calculator to graph two quadratic functions. They will describe and analyze the characteristics of both graphs, including the vertex.

Answers

1. parabola
2. upward
3. x^2, because it has a positive coefficient
4. table of (x, y) values that satisfy the function
5. Find the y-value where $x = 0$. This point is the vertex; (0, 4)
6. parabola
7. upward
8. x^2, because it has a positive coefficient
9. table of (x, y) values for both functions
10. Find the y-value where $x = 0$. This point is the vertex; (0, –4)
11. The value of the constant determines the vertex of the graph.

Station 3

Students will be given a graphing calculator. Students will use the graphing calculator to graph three quadratic functions. The quadratic functions have the same vertex, but varying widths. Students will describe the relationships between the widths of the parabolas.

Answers

1. vertex at (0, 0) because there is no constant
2. $y = x^2$; because the coefficient is 1 versus 3. A smaller coefficient yields a wider parabola.
3. table of x- and y-values for both graphs
4. $Y_2 = 3Y_1$; it shows that the y-value for each x-value of Y_2 is three times as large as Y_1.
5. because it has a smaller coefficient
6. table of x- and y-values for all three graphs
7. $Y_3 = 1/2(Y_1)$; it shows that the y-value for each x-value of Y_3 is half the size of Y_1.
8. $Y_3 = 1/6(Y_2)$; it shows that the y-value for each x-value of Y_3 is one-sixth the size of Y_2.

Station 4

Students will be given graph paper and a ruler. Students will analyze and graph quadratic equations using vertices, x-intercepts, and a table of values. They will determine why certain parabolas open upward while others open downward.

Answers

1. $f(x) = x^2 - x - 6$ and $f(x) = -x^2 + x - 6$

   $a = 1$ $\quad$ $a = -1$

   $b = -1$ $\quad$ $b = 1$

   $c = -6$ $\quad$ $c = -6$

2. For $f(x) = x^2 - x - 6$, the vertex is (1/2, –23/4).

   For $f(x) = -x^2 + x - 6$, the vertex is (1/2, –25/4).

3. (3, 0), (–2, 0)

4.

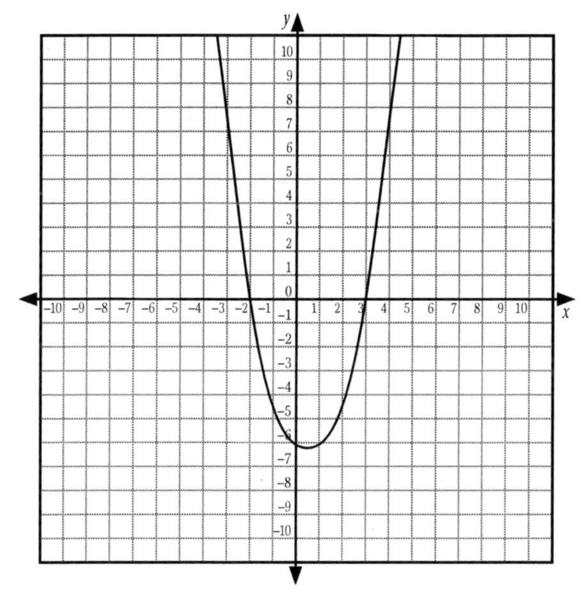

5. upward, because the coefficient of x^2 is positive

6. Table:

x	$y = f(x)$
–4	–26
0	–6
4	–18

Graph:

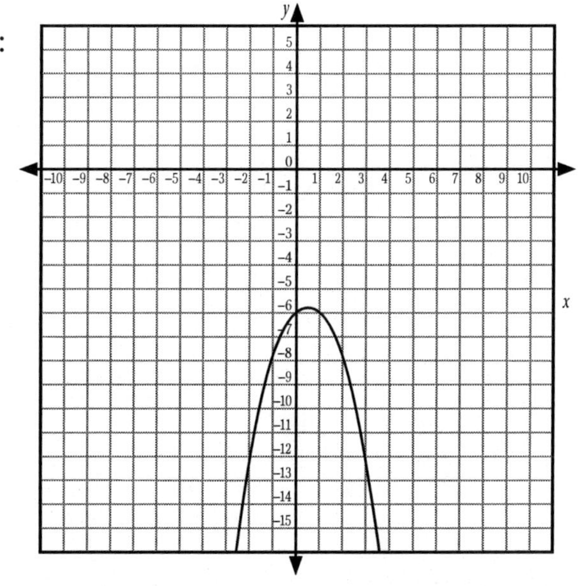

7. downward, because the coefficient of x^2 is negative

8. no, because the parabola opens downward

Materials List/Setup

Station 1 graph paper; ruler

Station 2 graphing calculator

Station 3 graphing calculator

Station 4 graph paper; ruler

Discussion Guide

To support students in reflecting on the activities and to gather some formative information about student learning, use the following prompts to facilitate a class discussion to "debrief" the station activities.

Prompts/Questions

1. What does the graph of a quadratic equation look like?
2. Why does the graph have this shape?
3. How do you find the vertex of a quadratic equation?
4. Does a quadratic equation with a term of x^2 open upward or downward? Why?
5. Does a quadratic equation with a term of $-x^2$ open upward or downward? Why?
6. How can you find the x-intercepts of a quadratic equation?

Think, Pair, Share

Have students jot down their own responses to questions, then discuss with a partner (who was not in their station group), and then discuss as a whole class.

Suggested Appropriate Responses

1. The graph of a quadratic equation looks like a parabola.
2. There are two x-values for each y-value because of the x^2 term.
3. vertex = $\left(\frac{-b}{2a}, f\left(\frac{-b}{2a}\right)\right)$
4. upward, because the coefficient is positive
5. downward, because the coefficient is negative
6. Use the vertex as the axis of symmetry and a table of values to determine the corresponding y-value when $x = 0$.

Possible Misunderstandings/Mistakes

- Not realizing that graphs of quadratic equations are parabolas
- Not using enough data points to construct a parabola for the graph of the quadratic equation
- Ignoring the sign of the coefficient of the x^2 term when constructing a parabola that opens upward or downward

Station 1

At this station, you will find graph paper and a ruler. Work together to graph the following quadratic equation:

$$y = x^2 + 6x + 9$$

1. Write this quadratic equation as a quadratic function. ____________________

2. What are the values of a, b, and c in the quadratic function?

   a = _______

   b = _______

   c = _______

To graph the function, you need the vertex, x-intercept, and y-intercept.

3. If the x-value of the vertex is found by $x = \frac{-6}{2(1)} = -3$, then write this x calculation using the general terms a, b, and/or c.

   ____

4. If the y-value of the vertex is found by $y = f\left(\frac{-6}{2(1)}\right) = f(-3) = 0$, then write this y calculation using the general terms a, b, and/or c.

   ____

5. Based on problems 3 and 4, how can you find the vertex of the graph for $f(x) = ax^2 + bx + c$?

   What is the vertex of the quadratic function $x^2 + 6x + 9 = 0$? ____________________

continued

6. How do you find the x-intercept of a function? (*Hint*: $y = f(x)$)

7. How do you find the y-intercept of a function?

8. What are the intercepts for $y = x^2 + 6x + 9$? ____________________

9. On your graph paper, graph the function using the vertex, x-intercept, and y-intercept.

10. What shape is the graph? Why do you think the graph has this shape?

Station 2

At this station, you will find a graphing calculator. As a group, follow the steps according to your calculator model to graph $y = x^2 + 4$ and $y = x^2 - 4$.

On a TI-83/84:	**On a TI-Nspire:**
Step 1: Press [Y=]. At Y_1, type [X,T,θ,n][x^2][+][4]. Step 2: Press [GRAPH].	Step 1: Arrow over to the graphing icon and press [enter]. At *f1*(*x*), enter [*x*], hit the [x^2] key, then type [+][4]. Step 2: Press [enter].

1. What shape is the graph? ____________________

2. Does the graph open upward or downward? ____________________

3. Which term do you think makes the graph open upward or downward? Explain your reasoning.

On a TI-83/84:	**On a TI-Nspire:**
Step 3: Press [2ND], then [GRAPH].	Step 3: Press [ctrl], then [T].

4. What information does your calculator show?

5. How can you use this information to find the vertex of the graph?

   What is the vertex of the graph? ____________________

continued

NAME: ___

Functions

Set 1: Graphing Quadratic Equations

On a TI-83/84:

Step 4: Press [Y=]. At Y_2, type [X,T,θ,n][x^2] [–][4].

Step 5: Press [GRAPH].

On a TI-Nspire:

Step 4: Press [ctrl][tab] to go back to the graphing window. Use the touch pad to select “>>” on the bottom left of the screen. At *f2*(*x*), enter [*x*], hit the [x^2] key, then type [–][4].

Step 5: Press [enter].

6. What shape is the graph? ___

7. Does the graph open upward or downward? ___

8. Which term do you think makes the graph open upward or downward? Explain your reasoning.

On a TI-83/84:

Step 6: Press [2ND], then [GRAPH].

On a TI-Nspire:

Step 6: Press [ctrl], then [T]. Press [ctrl], then [T] a second time to refresh the screen.

9. What information does your calculator show?

10. How can you use this information to find the vertex of the graph of $y = x^2 - 4$?

    What is the vertex of $y = x^2 - 4$? ___

11. Why do the graphs for $y = x^2 + 4$ and $y = x^2 - 4$ have different vertices?

Station 3

At this station, you will find a graphing calculator. As a group, follow the steps according to your calculator model to graph $y = x^2$, $y = 3x^2$, and $y = \frac{1}{2}x^2$.

On a TI-83/84:

Step 1: Press [Y=]. At Y_1, type [X,T,θ,n][x^2]. At Y_2, type [3][X,T,θ,n][x^2].

Step 2: Press [GRAPH].

On a TI-Nspire:

Step 1: Arrow over to the graphing icon and press [enter]. At *f1*(*x*), enter [*x*], then hit the [x^2] key. Arrow down. At *f2*(*x*), enter [3][*x*], then hit the [x^2] key.

Step 2: Press [enter].

1. Why do both graphs have the same vertex?

2. Which graph is wider, $y = x^2$ or $y = 3x^2$? ____________________

   Why is one graph wider than the other?

On a TI-83/84:

Step 3: Press [2ND], then [GRAPH].

On a TI-Nspire:

Step 3: Press [ctrl], then [T].

3. What information does your calculator show?

4. What is the relationship between Y1 and Y2 in the table?

   How does this relationship relate to $y = x^2$ and $y = 3x^2$?

continued

On a TI-83/84:

Step 4: Press [Y=]. At Y_3, type [0][.][5] [X,T,θ,n][x^2].

Step 5: Press [GRAPH].

On a TI-Nspire:

Step 4: Press [ctrl][tab] to go back to the graphing window. Use the touch pad to select ">>" on the bottom left of the screen. At $f3(x)$, enter [0][.][5][x], then hit the [x^2] key.

Step 5: Press [enter].

5. Why is the graph of $y = 0.5x^2$ wider than $y = x^2$ and $y = 3x^2$?

On a TI-83/84:

Step 6: Press [2ND], then [GRAPH].

On a TI-Nspire:

Step 6: Press [ctrl], then [T]. Press [ctrl], then [T] a second time to refresh the screen.

6. What information does your calculator show?

7. What is the relationship between Y1 and Y3 in the table?

   How does this relationship relate to $y = x^2$ and $y = 0.5x^2$?

8. What is the relationship between Y2 and Y3 in the table?

   How does this relationship relate to $y = 3x^2$ and $y = 0.5x^2$?

Station 4

At this station, you will find graph paper and a ruler. Work together to graph the following quadratic equations:

$$f(x) = x^2 - x - 6 \text{ and } f(x) = -x^2 + x - 6$$

1. What are the values of a, b, and c in each quadratic function?

$f(x) = x^2 - x - 6$	$f(x) = -x^2 + x - 6$
$a =$ _______	$a =$ _______
$b =$ _______	$b =$ _______
$c =$ _______	$c =$ _______

2. Use the information in problem 1 to find the vertex $\left(\frac{-b}{2a}, f\left(\frac{-b}{2a}\right)\right)$ for each function. Show your work.

3. Find the x-intercepts of $f(x) = x^2 - x - 6$ using factoring. Show your work.

4. On your graph paper, graph $f(x) = x^2 - x - 6$ using its vertex and x-intercepts.

5. Does the parabola open upward or downward? Explain your answer.

continued

6. Fill out the table below to help you graph $f(x) = -x^2 + x - 6$.

x	$y = f(x)$
–4	
0	
4	

Graph $f(x) = -x^2 + x - 6$ on your graph paper.

7. Does the graph open upward or downward? Explain your answer.

8. Will the graph of $f(x) = -x^2 + x - 6$ have x-intercepts? Why or why not?

Congruence

Set 1: Parallel Lines and Transversals

Instruction

Goal: To provide opportunities for students to develop concepts and skills related to identifying and using the relationships between special pairs of angles formed by parallel lines and transversals

Common Core State Standard

G–CO.9 Prove theorems about lines and angles. *Theorems include: vertical angles are congruent; when a transversal crosses parallel lines, alternate interior angles are congruent and corresponding angles are congruent; points on a perpendicular bisector of a line segment are exactly those equidistant from the segment's endpoints.*

Station 1

Students will be given graph paper, a ruler, and a protractor. Students will use the graph paper and ruler to model parallel lines cut by a transversal. They will use the protractor to find vertical angles. Then they will use the graph paper to model lines, which are not parallel, that are cut by a transversal. They will use the protractor to find vertical angles. They will realize that vertical angles can be found by both parallel and non-parallel lines cut by a transversal.

Answers

1. Answers will vary.
2. $\angle A \cong \angle C$; $\angle B \cong \angle D$
3. vertical angles
4. $\angle E \cong \angle H$; $\angle F \cong \angle G$
5. Answers will vary.
6. Answers will vary.
7. $\angle E \cong \angle H$; $\angle F \cong \angle G$
8. vertical angles
9. $\angle A \cong \angle D$; $\angle B \cong \angle C$
10. Answers will vary.
11. Both parallel and non-parallel lines have vertical angles when cut by a transversal.

Station 2

Students will be given graph paper, a ruler, and a protractor. Students will use the graph paper and ruler to model parallel lines cut by a transversal. They will use the protractor to find supplementary angles. Then they will use the graph paper to model lines, which are not parallel, that are cut by a transversal. They will use the protractor to find supplementary angles. They will describe which types of angles are supplementary angles when two lines are cut by a transversal.

Answers

1. $\angle A$ and $\angle B$; $\angle C$ and $\angle D$; $\angle E$ and $\angle F$; $\angle G$ and $\angle H$; $\angle A$ and $\angle C$; $\angle B$ and $\angle D$; $\angle E$ and $\angle G$; $\angle F$ and $\angle H$
2. Answers will vary.
3. $\angle A$ and $\angle B$; $\angle C$ and $\angle D$; $\angle E$ and $\angle F$; $\angle G$ and $\angle H$; $\angle A$ and $\angle C$; $\angle B$ and $\angle D$; $\angle E$ and $\angle G$; $\angle F$ and $\angle H$
4. Answers will vary.
5. adjacent angles; interior angles on the same side of the transversal but only when the transversal intersects parallel lines.

Station 3

Students will be given spaghetti noodles, a protractor, graph paper, and a ruler. Students will use the graph paper and spaghetti noodles to model parallel lines cut by a transversal. They will use the protractor to measure the angles created by the transversal. Then they will explore and define the exterior, interior, alternate exterior, and alternate interior angles created by the transversal.

Answers

1. 8 angles
2. interior
3. exterior
4. 1, 2, 7, 8
5. 3, 4, 5, 6
6. Answers will vary.
7. $m\angle 1 = m\angle 7$; $m\angle 2 = m\angle 8$; $m\angle 3 = m\angle 5$; $m\angle 4 = m\angle 6$
8. 4 and 6; 3 and 5
9. 1 and 7; 2 and 8
10. equal; equal

Station 4

Students will be given spaghetti noodles, a protractor, and two parallel lines cut by a transversal. Students will use the spaghetti noodles to model the letter "F" to find corresponding angles. Then they will use the protractor to measure the angles and justify their answer.

Answers

1. 3 and 7
2. 1 and 5; 2 and 6; 3 and 7; 4 and 8
3. Answers will vary.
4. Answers will vary.

Materials List/Setup

Station 1 graph paper; ruler; protractor

Station 2 graph paper; ruler; protractor

Station 3 dry spaghetti noodles; protractor; graph paper; ruler

Station 4 dry spaghetti noodles; protractor

Discussion Guide

To support students in reflecting on the activities and to gather some formative information about student learning, use the following prompts to facilitate a class discussion to "debrief" the station activities.

Prompts/Questions

1. How many angles are created when parallel or non-parallel lines are cut by a transversal?
2. What are exterior angles?
3. What are interior angles?
4. What is a vertical angle?
5. What are alternate exterior angles?
6. What are alternate interior angles?
7. What two types of angles are supplementary when parallel lines are cut by a transversal?

Think, Pair, Share

Have students jot down their own responses to questions, then discuss with a partner (who was not in their station group), and then discuss as a whole class.

Suggested Appropriate Responses

1. 8 angles
2. Exterior angles lie on the outside of the lines cut by the transversal.
3. Interior angles lie in between the two lines cut by the transversal.
4. Vertical angles are two angles formed by two intersecting lines that lie on opposite sides of the point of intersection.
5. Alternate exterior angles are pairs of angles on opposite sides of the transversal that are outside of the parallel or non-parallel lines.
6. Alternate interior angles are pairs of angles on opposite sides of the transversal that are inside the parallel or non-parallel lines.
7. Interior angles on the same side of the transversal and adjacent angles are supplementary in this situation.

Possible Misunderstandings/Mistakes

- Not finding corresponding angles correctly and identifying that they have the same measure if the lines cut by the transversal are parallel
- Mixing up interior and exterior angles
- Not realizing vertical angles are always equal whether or not the lines cut by the transversal are parallel
- Not realizing that alternate interior or exterior angles must be on the opposite side of the transversals

NAME: ______________________________

Congruence

Set 1: Parallel Lines and Transversals

Station 1

At this station, you will find graph paper, a ruler, and a protractor. As a group, construct two parallel lines that are cut by a transversal and label the angles as shown in the diagram below.

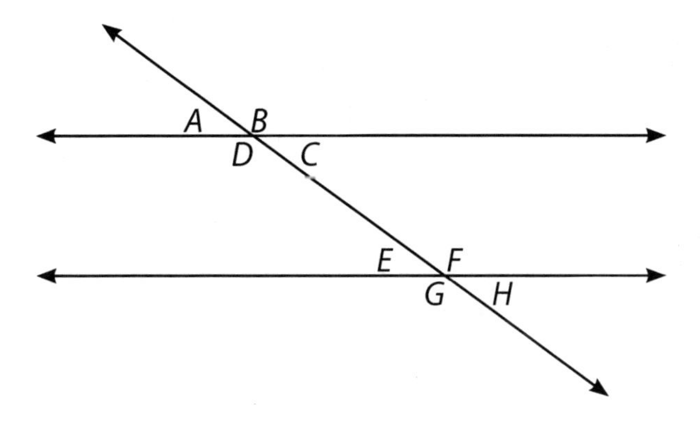

1. Use the protractor to measure the following angles:

   $m\angle A =$ ______ $m\angle B =$ ______ $m\angle C =$ ______ $m\angle D =$ ______

2. Which pairs of angles are equal in problem 1?

   ____

3. The pairs of angles found in problem 2 have a special name. What are these angles called? (*Hint:* Think about their location in relation to each other.)

   ____

4. What pairs of angles are congruent for angles *E*, *F*, *G*, and *H*?

   ____

5. What strategy did you use to find the angle pairs in problem 4?

continued

Congruence

Set 1: Parallel Lines and Transversals

On your graph paper, construct a new graph of two lines that are NOT parallel. These lines are cut by a transversal. Label the angles as shown in the diagram below.

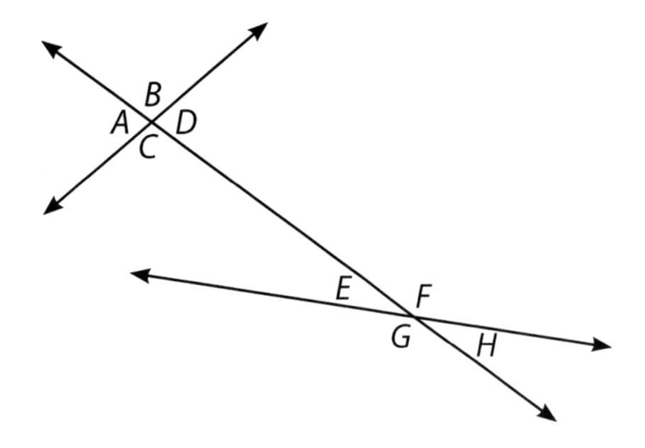

6. Use the protractor to measure the following angles:

   $m\angle E$ = ______ $m\angle F$ = ______ $m\angle G$ = ______ $m\angle H$ = ______

7. Which pairs of angles are equal?

8. The pairs of angles found in problem 7 have a special name. What are these angles called? (*Hint:* Think about their location in relation to each other.)

   ____

9. What pairs of angles are congruent for angles *A*, *B*, *C*, and *D*?

   ____

10. What strategy did you use to find the angle pairs in problem 9?

11. Based on your observations in problems 1–10, vertical angles can be found for what types of lines cut by a transversal?

NAME:

Congruence

Set 1: Parallel Lines and Transversals

Station 2

At this station, you will find graph paper, a ruler, and a protractor. As a group, construct two parallel lines that are cut by a transversal and label the angles as shown in the diagram below.

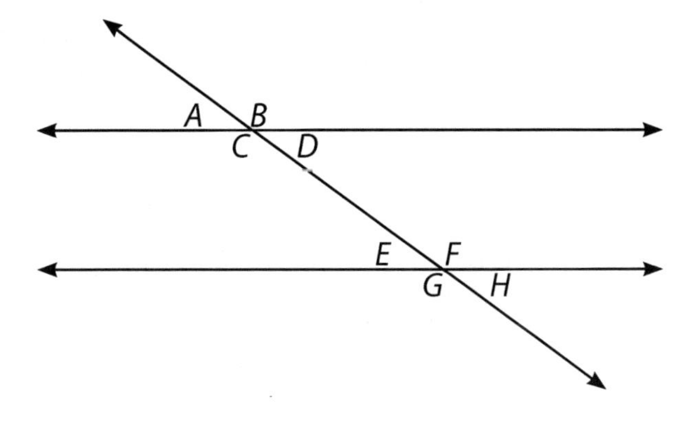

1. Which angles are supplementary angles?

2. Use your protractor to justify your answer to problem 1 by recording the measurements of each angle below.

$m\angle A =$ ______ $m\angle C =$ ______ $m\angle E =$ ______ $m\angle G =$ ______

$m\angle B =$ ______ $m\angle D =$ ______ $m\angle F =$ ______ $m\angle H =$ ______

On your graph paper, construct a new graph of two lines that are NOT parallel. These lines are cut by a transversal. Label the angles as shown in the diagram below.

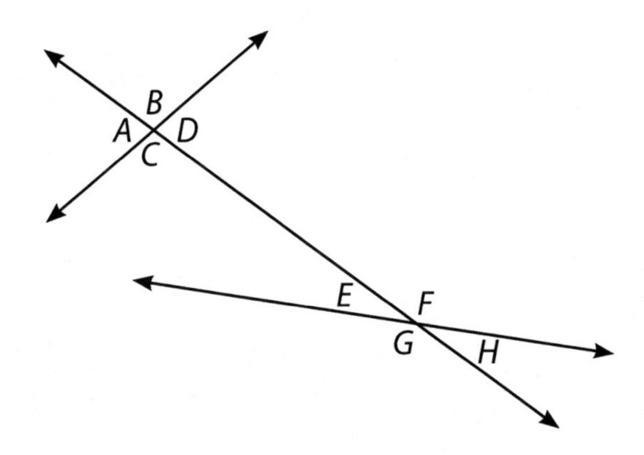

3. Which angles are supplementary angles?

continued

4. Use your protractor to justify your answer to problem 3 by recording the measurements of each angle below.

   $m\angle A =$ ______ $m\angle C =$ ______ $m\angle E =$ ______ $m\angle G =$ ______

   $m\angle B =$ ______ $m\angle D =$ ______ $m\angle F =$ ______ $m\angle H =$ ______

5. Based on your answers and observations in problems 1–4, which of the following types of angles are supplementary?

   vertical angles adjacent angles alternate interior angles

   alternate exterior angles interior angles on the same side of the transversal

NAME: ____________________

Congruence

Set 1: Parallel Lines and Transversals

Station 3

At this station, you will find spaghetti noodles, a protractor, graph paper, and a ruler. Follow the directions below, and then answer the questions.

- On the graph paper, construct two parallel horizontal lines.
- Construct a diagonal line that passes through both parallel lines. This line is called a transversal.

1. How many angles does this transversal create with the two parallel lines?

   ____

On your graph paper, label the angles to model the parallel lines and transversal shown below.

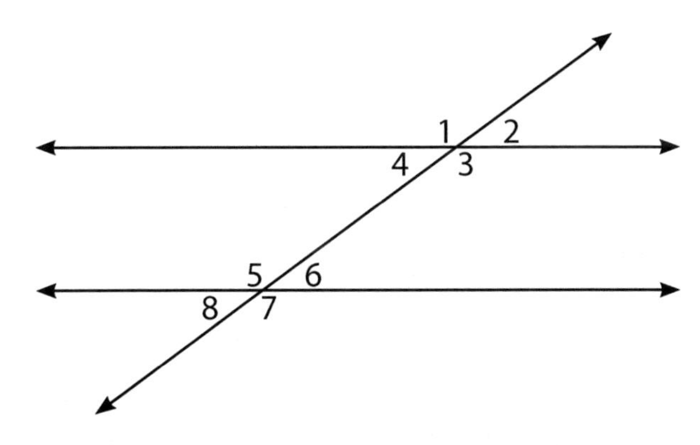

Place a spaghetti noodle on each parallel line.

2. Are the angles between the spaghetti noodles called exterior or interior angles?

   ____

3. Are the angles above the top parallel spaghetti noodle and below the bottom spaghetti noodle called exterior or interior angles?

   ____

continued

4. Which angles are exterior angles? ____________________

5. Which angles are interior angles? ____________________

6. Use your protractor to measure each angle. Write the angle measurements below.

   $m\angle 1 =$ _______ $m\angle 3 =$ _______ $m\angle 5 =$ _______ $m\angle 7 =$ _______

   $m\angle 2 =$ _______ $m\angle 4 =$ _______ $m\angle 6 =$ _______ $m\angle 8 =$ _______

7. Based on your answers in problem 6, which pairs of interior and which pairs of exterior angles are equal?

8. Based on your answers in problems 6 and 7, which angles are alternate interior angles?

9. Based on your answers in problems 6 and 7, which angles are alternate exterior angles?

10. Based on your observations, alternate interior angles have ____________________ measure.

    Alternate exterior angles have ____________________ measure.

NAME: ____________________

Congruence

Set 1: Parallel Lines and Transversals

Station 4

At this station, you will find spaghetti noodles, a protractor, and parallel lines cut by a transversal.

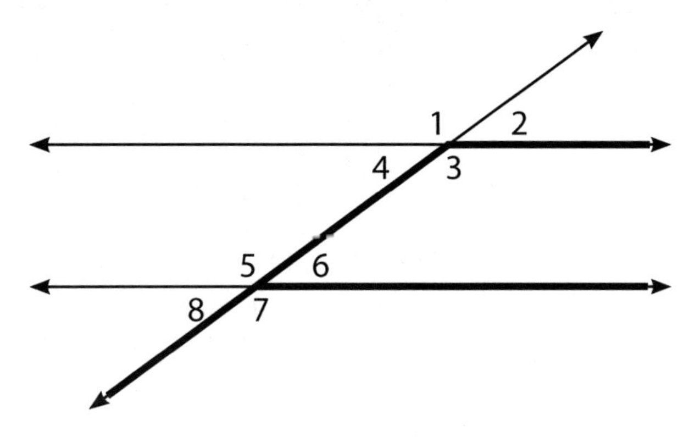

The letter "F" has been drawn on the diagram above in a bold line. Place spaghetti noodles on the "F".

1. What angles are at the inside corners of the "F"? ____________________

   These are called *corresponding angles*.

2. Move the spaghetti noodles around to create more "F's" that help you find the corresponding angles. Write the pairs of corresponding angles in the space below.

3. What strategy did you use to move the letter "F" around to help you find the corresponding angles?

4. Use your protractor to measure each angle to justify your answers for problem 2.

   $m\angle 1$ = ______ $m\angle 3$ = ______ $m\angle 5$ = ______ $m\angle 7$ = ______

   $m\angle 2$ = ______ $m\angle 4$ = ______ $m\angle 6$ = ______ $m\angle 8$ = ______

Congruence

Set 2: Rhombi, Squares, Kites, and Trapezoids

Goal: To provide opportunities for students to develop concepts and skills related to describing and comparing relationships among quadrilaterals, including squares, rectangles, rhombi, parallelograms, trapezoids, and kites

Common Core State Standard

G–CO.11 Prove theorems about parallelograms. *Theorems include: opposite sides are congruent, opposite angles are congruent, the diagonals of a parallelogram bisect each other, and conversely, rectangles are parallelograms with congruent diagonals.*

Student Activities Overview and Answer Key

Station 1

Students will be given four drinking straws, tape, a ruler, and a protractor. Students will construct a square. Then they will "lean" the square to create a rhombus. They describe why a rhombus is a parallelogram. Then they relate a square to a rhombus.

Answers

1. 90°
2. yes
3. yes
4. yes
5. yes
6. rhombus
7. A rhombus is like a square because all four sides are congruent.

Station 2

Students will be given a ruler and a protractor. Students will construct a square. They will find the diagonals of the square. Then they will relate the square to a rectangle, a rhombus, and a parallelogram.

Answers

1. 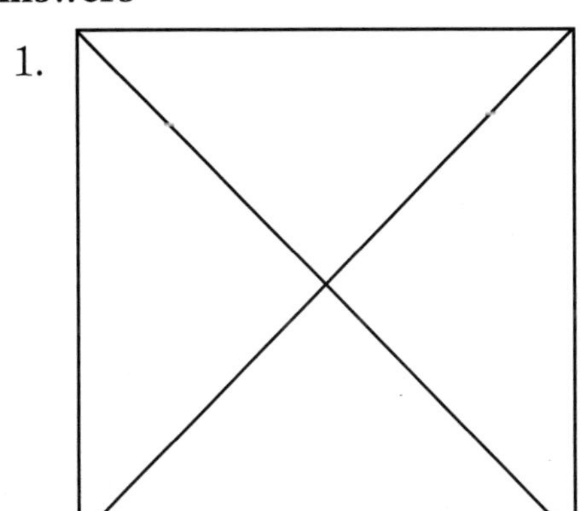
2. by using the Pythagorean Theorem
3. $2\sqrt{2}$
4. A square is always a rectangle. A square is a rectangle with four equal sides.
5. A square is a rhombus with four equal angles.
6. A square is a parallelogram with equal diagonals that bisect the angles.

Station 3

Students will be given three drinking straws, a ruler, tape, and scissors. Students will create a kite from the drinking straws, and describe the sides, angles, and diagonals in the kite. They will determine if a kite is a parallelogram.

Answers

1. Answers will vary.
2. Answers will vary.
3. Answers will vary.
4. Two pairs of adjacent sides are congruent.
5. Draw the horizontal diagonal to make the base of each isosceles triangle.
6. A kite is not a parallelogram because opposite sides are not congruent.
7. Diagonals are perpendicular.

Station 4

Students will be given tracing paper, a ruler, and scissors. They will construct and describe a trapezoid, and explain if the trapezoid can be turned into a parallelogram. They will construct and describe an isosceles trapezoid. They will explain if the trapezoid can be turned into a parallelogram.

Answers

1. Yes, because it has four sides.
2. No, because opposite sides are not congruent.
3. No, because the sides of the trapezoid consist of two non-congruent triangles and a rectangle.
4. Yes, because it has four sides.
5. No, because opposite sides are not congruent.
6. Yes, because the trapezoid consists of two congruent isosceles triangles and a rectangle; rectangle

Materials List/Setup

Station 1 four drinking straws; tape; ruler; protractor

Station 2 ruler; protractor

Station 3 three drinking straws; ruler; tape; scissors

Station 4 tracing paper; ruler; scissors

Discussion Guide

To support students in reflecting on the activities and to gather some formative information about student learning, use the following prompts to facilitate a class discussion to "debrief" the station activities.

Prompts/Questions

1. What is a rhombus? Is it a parallelogram?
2. How does a square relate to a rhombus?
3. What is a kite? Is it a parallelogram?
4. What is a trapezoid? Is it a parallelogram?

Think, Pair, Share

Have students jot down their own responses to questions, then discuss with a partner (who was not in their station group), and then discuss as a whole class.

Suggested Appropriate Responses

1. A rhombus is a quadrilateral with all four sides equal and opposite angles equal. Yes, it is a parallelogram.
2. A square is a rhombus with four right angles.
3. A kite is a quadrilateral that has two pairs of adjacent sides that are equal. No, it is not a parallelogram.
4. A trapezoid is a quadrilateral that has one pair of parallel sides. No, it is not a parallelogram.

Possible Misunderstandings/Mistakes

- Not recognizing that a square is a type of rhombus
- Not recognizing that all sides are equal in a rhombus
- Not realizing that all squares are rectangles, but not all rectangles are squares
- Not recognizing that pairs of adjacent sides in a kite have equal length

Station 1

At this station, you will find four drinking straws, tape, a ruler, and a protractor. Work together to answer the questions.

1. Tape the four drinking straws together to create a square. What are the measures of each interior angle?

2. "Lean" the straws so that you create interior angles of 120°, 60°, 120°, and 60°. Are the lengths of the sides still equal? Why or why not?

3. Are pairs of opposite sides congruent?

4. Are pairs of opposite angles congruent?

5. Is this figure a parallelogram? Why or why not?

6. What is the special name for this quadrilateral?

7. How does this quadrilateral relate to a square?

Station 2

At this station, you will find a ruler and a protractor. Work together to construct the figures and answer the questions.

1. In the space below, construct a square with side lengths of 2 inches. Construct the diagonals on the square.

2. How can you find the length of each diagonal without using your ruler?

3. What is the length of each diagonal? Verify your answer with your ruler.

4. Is a square always a rectangle or is a rectangle always a square? Justify your answer.

5. How can you relate a square to a rhombus?

continued

In the space below, draw a square and rhombus to show how they relate to each other.

6. How can you relate a square to a parallelogram?

Station 3

At this station, you will find three drinking straws, a ruler, tape, and scissors. Work as a group to create a kite.

1. Cut one straw exactly in half. Tape the four straws together to create a kite similar to the image below.

2. What is the length of each side?

3. What is the measure of each angle?

4. What is the relationship between adjacent sides of a kite?

5. How can you create two isosceles triangles from the kite?

6. Is a kite a parallelogram? Why or why not?

continued

7. Place your straw kite in the space below. Draw the diagonals of the kite. What relationship do the diagonals have to each other?

Station 4

At this station, you will find tracing paper, a ruler, and scissors. Work as a group to answer the questions.

Given the following trapezoid:

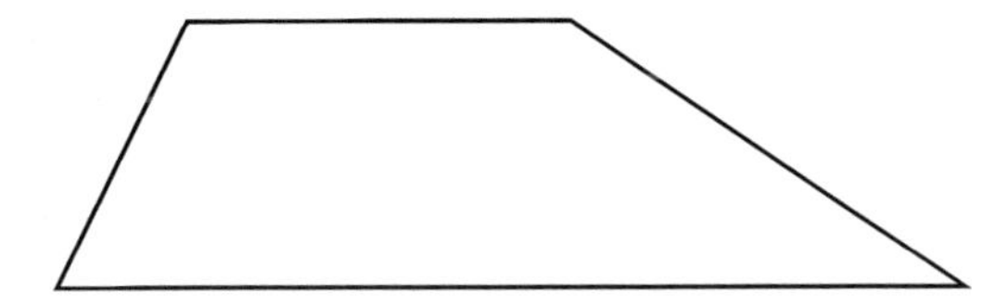

Use the tracing paper to trace the trapezoid above. Cut the trapezoid out of the tracing paper.

1. Is this trapezoid a quadrilateral? Why or why not?

2. Is this trapezoid a parallelogram? Why or why not?

   If so, what type of parallelogram did you create?

3. Can you cut the trapezoid into two pieces and create a parallelogram? Why or why not?

continued

Congruence

Set 2: Rhombi, Squares, Kites, and Trapezoids

Given the following trapezoid:

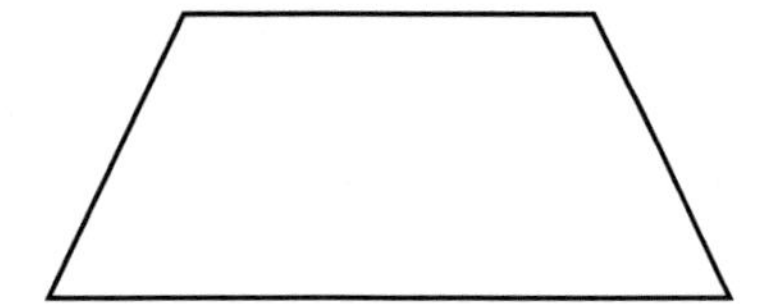

Use the tracing paper to trace the trapezoid above. Cut the trapezoid out of the tracing paper.

4. Is this trapezoid a quadrilateral? Why or why not?

5. Is this trapezoid a parallelogram? Why or why not?

6. Can you cut the trapezoid into two pieces and create a parallelogram? Why or why not?

   If so, what type of parallelogram did you create?

Congruence

Set 3: Circumcenter, Incenter, Orthocenter, and Centroid

Instruction

Goal: To provide opportunities for students to develop concepts and skills related to the circumcenter, incenter, orthocenter, and centroid of a circle

Common Core State Standards

G–CO.12 Make formal geometric constructions with a variety of tools and methods (compass and straightedge, string, reflective devices, paper folding, dynamic geometric software, etc.). *Copying a segment; copying an angle; bisecting a segment; bisecting an angle; constructing perpendicular lines, including the perpendicular bisector of a line segment; and constructing a line parallel to a given line through a point not on the line.*

G–CO.13 Construct an equilateral triangle, a square, and a regular hexagon inscribed in a circle.

Student Activities Overview and Answer Key

Station 1

Students will be given a ruler, a protractor, and a compass. They will construct a square and circumscribe a circle around the square. Then they will construct the perpendicular bisectors of the square. They will realize that the circumcenter is the intersection of the perpendicular bisectors, which is also the center of the circle. They will repeat this process for an equilateral triangle.

Answers

1–2.

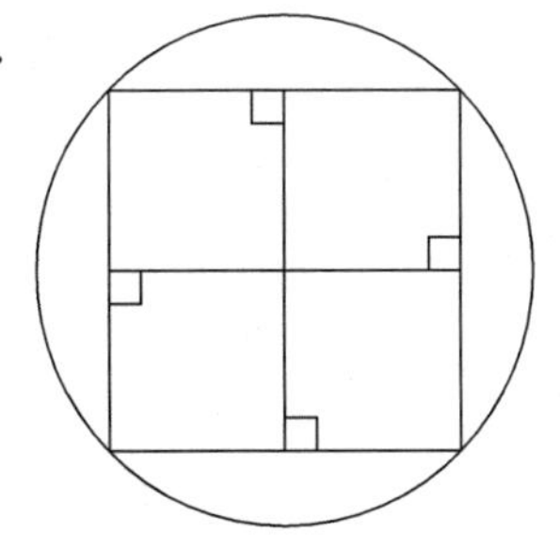

The circle circumscribes the square with the intersection of the perpendicular bisectors as the center of the circle.

3.

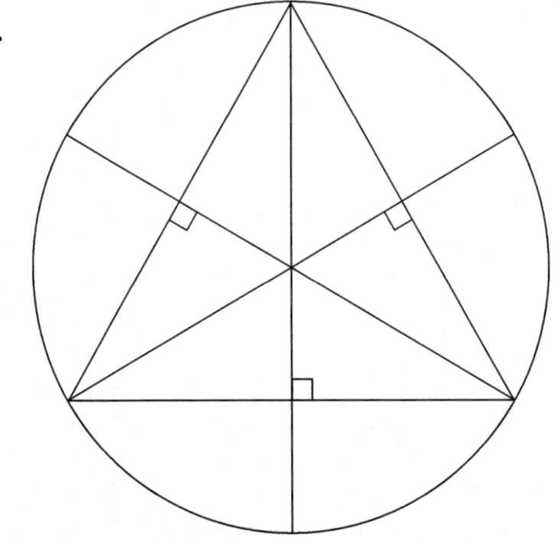

4. circumcenter
5. The circle circumscribes the triangle with each vertex being on the edge of the circle and the center being the intersection of the perpendicular bisectors.
6. The circumcenter is the center of the circle that circumscribes the polygon.

Station 2

Students will be given a ruler, a compass, and a protractor. They will construct the angle bisectors of an equilateral triangle. Then they will inscribe a circle in the triangle. They will realize that the incenter is the intersection of the angle bisectors, which is also the center of the circle.

Answers

1., 3., 4.

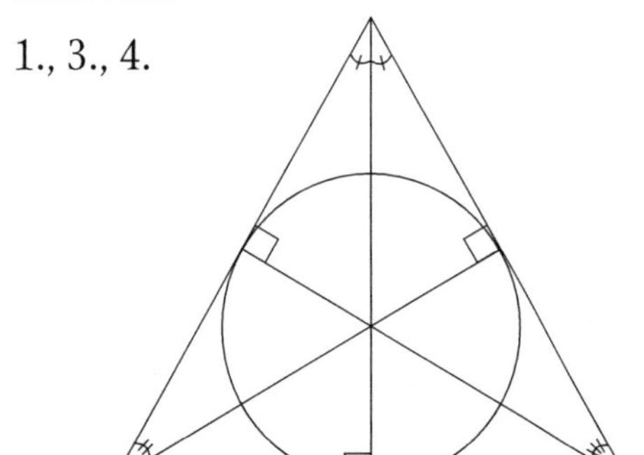

2. incenter
5. The circle is inscribed in the triangle.
6. The incenter is the center of an inscribed circle and the point where the angle bisectors of a regular polygon intersect.

Station 3

Students will be given a ruler, a protractor, and a compass. Students will construct the altitudes of an acute triangle to find the orthocenter. Then they will construct the altitudes of a right triangle and an obtuse triangle and discover that the orthocenter can lie inside, on, or outside the triangle.

Answers

1.

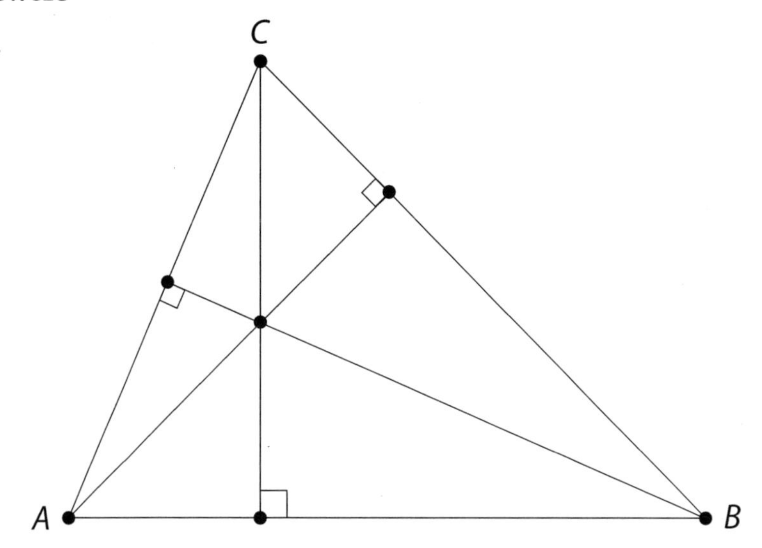

2. orthocenter
3. acute
4.

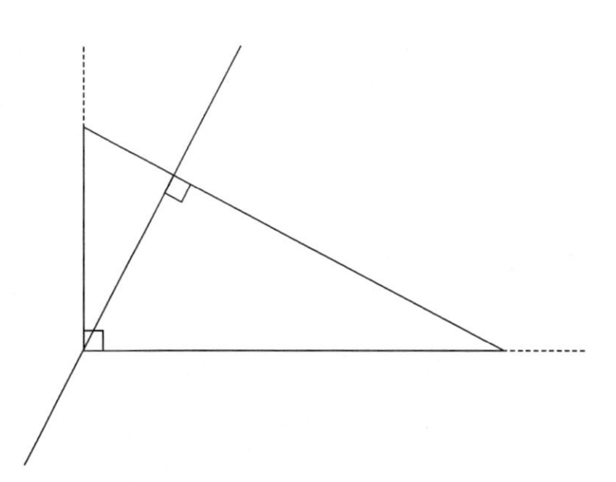

5. on the right angle of the triangle
6. right

7.

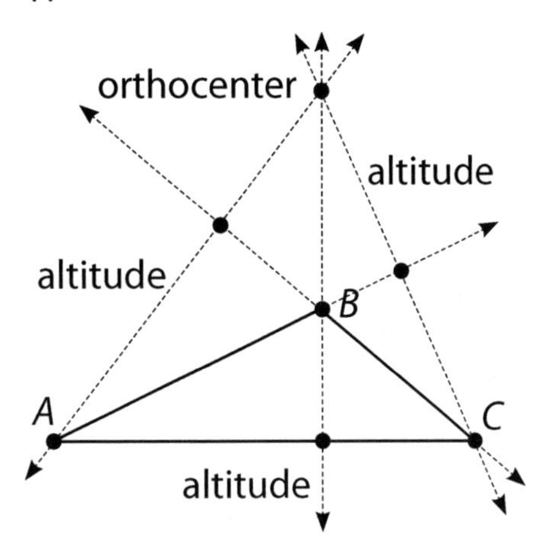

8. outside the triangle
9. obtuse
10. The orthocenter can lie inside, on, or outside the triangle.

Station 4

Students will be given notecards, scissors, a compass, a ruler, and a protractor. They will construct the centroid of three triangles. They will relate the centroid to the center of gravity or the balancing point.

Answers

1. Triangles may vary. Be sure all angles are acute in the triangle. Sample triangle with medians:

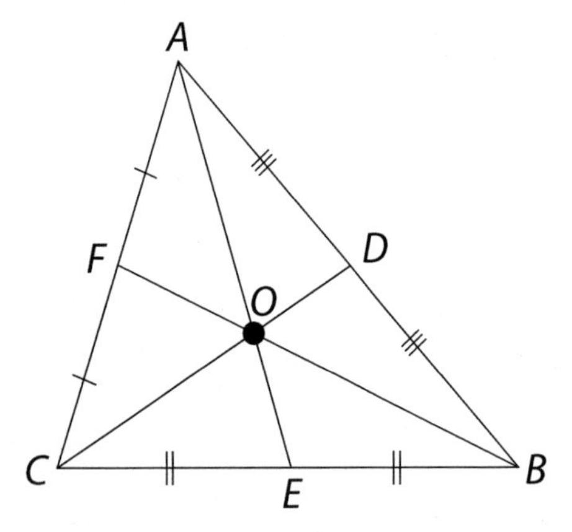

2. centroid

3. Triangles may vary. Be sure there is a right angle in the triangle. Sample triangle with medians:

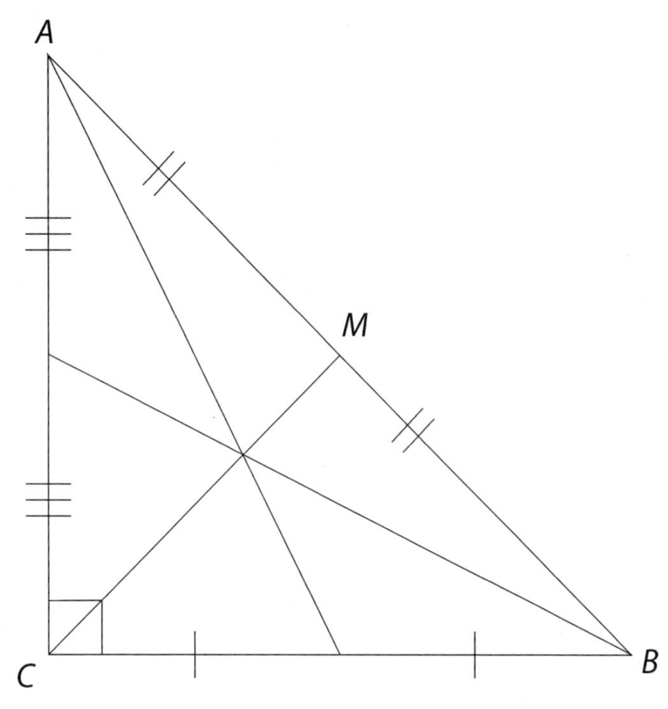

4. Triangles may vary. Be sure there is an obtuse angle in the triangle. Sample triangle with medians:

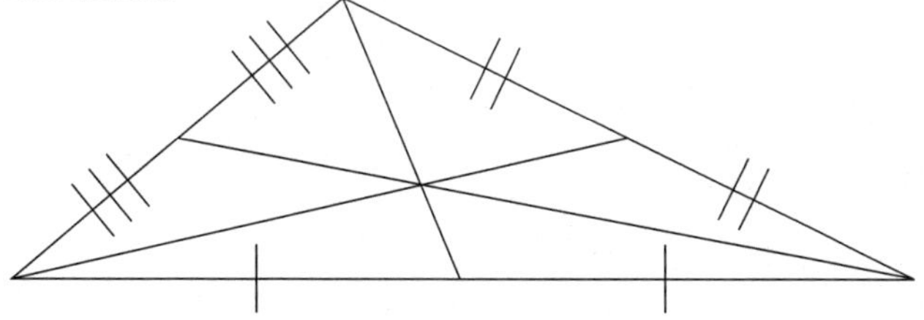

5. at the intersection of the medians
6. centroid
7. The centroid is the center of gravity and is formed by the intersection of the medians.

Materials List/Setup

Station 1 ruler; compass; protractor

Station 2 ruler; compass; protractor

Station 3 ruler; compass; protractor

Station 4 notecards; scissors; ruler; compass; protractor

Discussion Guide

To support students in reflecting on the activities and to gather some formative information about student learning, use the following prompts to facilitate a class discussion to "debrief" the station activities.

Prompts/Questions

1. What is the circumcenter of a circle circumscribed about a regular polygon?
2. What is the incenter of a circle inscribed in a regular polygon?
3. What is the orthocenter?
4. What is the centroid of a circle?

Think, Pair, Share

Have students jot down their own responses to questions, then discuss with a partner (who was not in their station group), and then discuss as a whole class.

Suggested Appropriate Responses

1. It is the intersection of the perpendicular bisectors of the circumscribed polygons.
2. It is the intersection of the angle bisectors of the inscribed polygons.
3. It is the intersection of the altitudes of a triangle.
4. The centroid is the intersection of the medians and the center of gravity for the figure.

Possible Misunderstandings/Mistakes

- Not realizing that the circumcenter is found in circumscribed circles
- Not realizing that the incenter is found in inscribed circles
- Incorrectly constructing perpendicular bisectors, angle bisectors, and altitudes

NAME:

Congruence

Set 3: Circumcenter, Incenter, Orthocenter, and Centroid

Station 1

At this station, you will find a ruler, a compass, and a protractor. Work as a group to answer the questions.

1. In the space below, construct the perpendicular bisectors of the sides of the square.

2. Draw a circle using the intersection of the perpendicular bisectors as the center of the circle and one of the vertices of the square as the radius. What do you notice about the circle in relation to the square?

continued

3. In the space below, construct the perpendicular bisectors of the sides of the triangle.

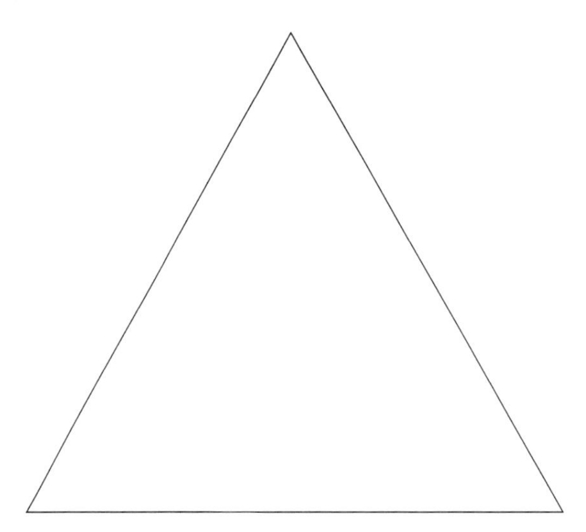

4. What is the name of the point of intersection of the perpendicular bisectors of a triangle?

5. Draw a circle using the intersection of the perpendicular bisectors as the center of the circle and one of the vertices of the triangle as the radius. What do you notice about the circle in relationship to the triangle?

6. Based on your observations in problems 1–4, what is the definition of the circumcenter of a circumscribed polygon?

Station 2

At this station, you will find a ruler, a compass, and a protractor. Work as a group to answer the questions.

1. Using the triangle below, bisect each angle of the triangle.

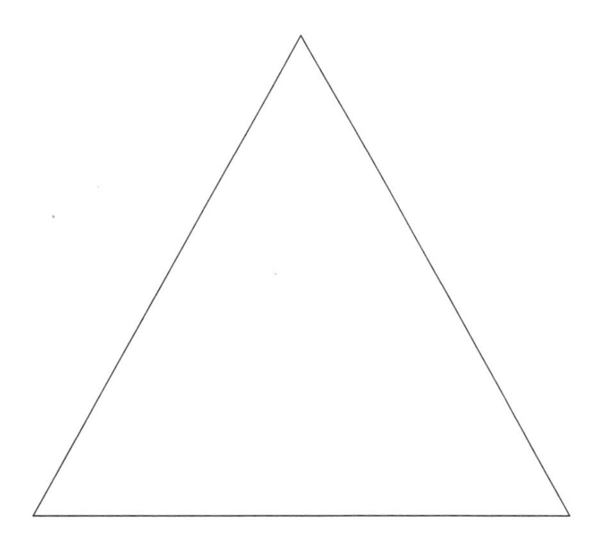

2. What is the name of the intersection point of the angle bisectors of a triangle?

3. Construct a perpendicular line from the intersection of the angle bisectors of the triangle to one of the sides of the triangle.

4. Draw a circle using the intersection of the angle bisectors of the triangle as the center and the intersection of the perpendicular line constructed in problem 2 with the side of the triangle as the radius.

continued

5. What do you notice about the circle in relation to the triangle?

6. Based on your observations in problems 1–5, what is the definition of the incenter of an inscribed polygon?

Station 3

At this station, you will find a ruler, a compass, and a protractor. Work as a group to answer the questions.

1. On the triangle below, construct the altitudes.

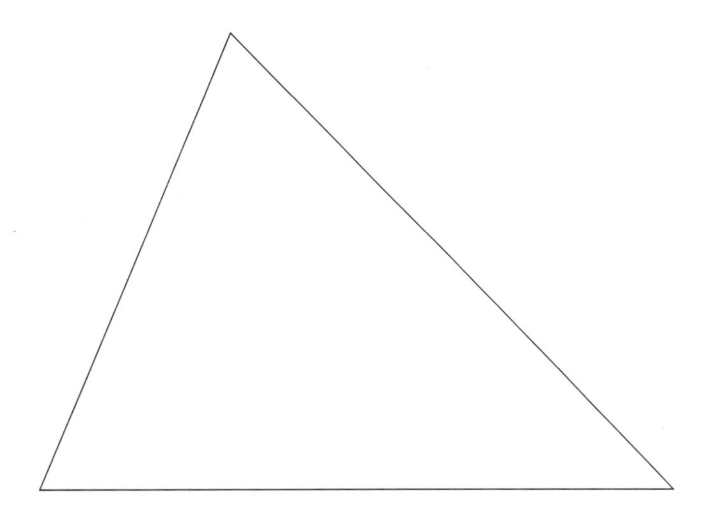

2. What is the name of the point of intersection of the altitudes of a triangle?

3. Based on the angles, what type of triangle is this?

continued

4. On the triangle below, construct the altitudes.

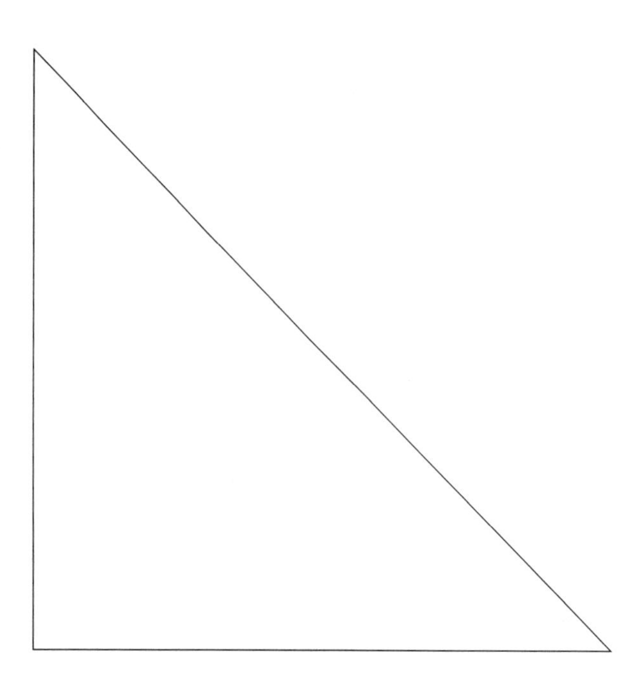

5. Where does the intersection of the altitudes occur?

6. Based on the angles, what type of triangle is this?

continued

7. On the triangle below, construct the altitudes. Remember that you might need to extend the leg(s) of the triangle to create the altitude(s).

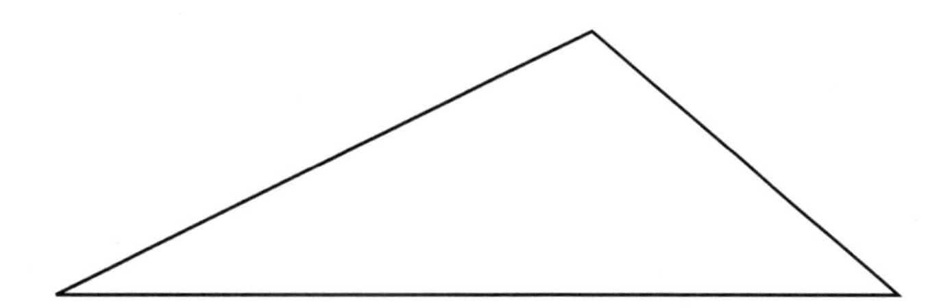

8. Where does the intersection of the altitudes occur?

9. Based on the angles, what type of triangle is this?

10. Based on your answers to problems 1–9, what conclusion can you draw about the location of the intersection of the altitudes of a triangle?

Station 4

At this station, you will find notecards, scissors, a compass, a ruler, and a protractor. Work together to construct the medians of the triangle and answer the questions.

1. Draw an acute triangle on one of the notecards and then construct the medians of the triangle.

2. Identify the intersection of the medians with a point. What is this point called?

3. Draw a right triangle on one of the notecards and then construct the medians of the triangle.

4. Draw an obtuse triangle on one of the notecards and then construct the medians of the triangle.

5. Cut out each of the triangles and try balancing each one of them on the tip of your finger. Where does the balancing point occur?

6. What is the intersection point of the medians called?

7. What conclusion can you draw about the intersection point of the medians of a triangle?

Similarity, Right Triangles, and Trigonometry

Set 1: Similarity and Scale Factor

Instruction

Goal: To provide opportunities for students to develop concepts and skills related to proving that triangles are similar using scale factors and angle relationships

Common Core State Standards

G–SRT.1 Verify experimentally the properties of dilations given by a center and a scale factor:

b. The dilation of a line segment is longer or shorter in the ratio given by the scale factor.

G–SRT.2 Given two figures, use the definition of similarity in terms of similarity transformations to decide if they are similar; explain using similarity transformations the meaning of similarity for triangles as the equality of all corresponding pairs of angles and the proportionality of all corresponding pairs of sides.

Student Activities Overview and Answer Key

Station 1

Students will be given graph paper, a ruler, a red marker, a green marker, and a blue marker. Students will construct similar and non-similar rectangles. They will determine ratios of corresponding sides to see if the rectangles are similar. Then they will find the relationship between perimeter and area with the corresponding sides of similar rectangles.

Answers

3. red rectangle:

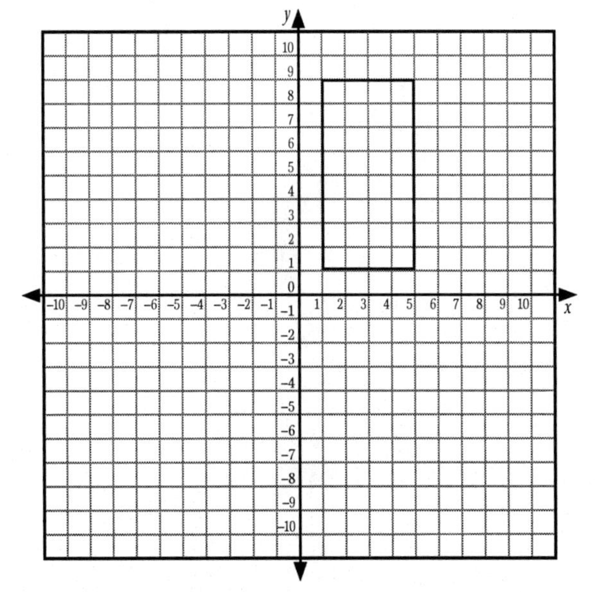

4. green rectangle:

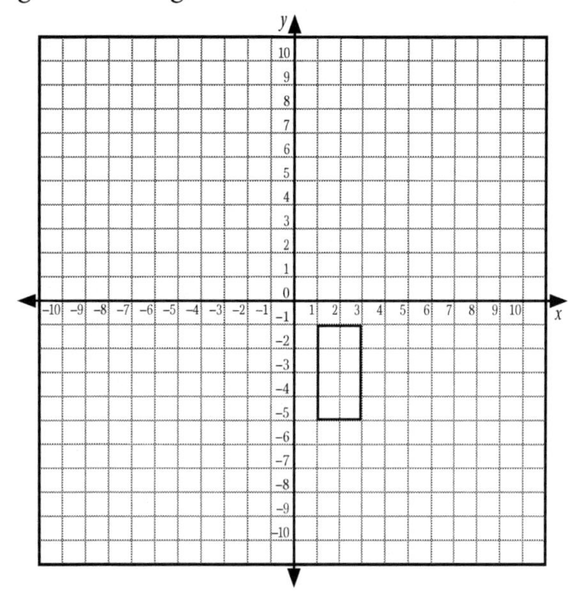

5. blue rectangle:

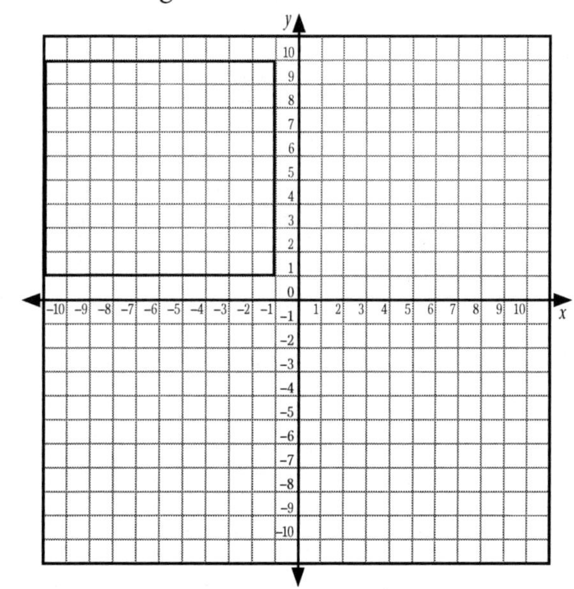

6. 4/2; 8/4; 4/2 = 8/4
7. 4/9; 8/10; 4/9 ≠ 8/10
8. red and green
9. red = 24 units, green = 12 units, blue = 38 units
10. red = 32 square units, green = 8 square units, blue = 90 square units

11. The perimeter has the same ratio as the corresponding sides.
12. The area is equal to the square of the ratio of the corresponding sides.

Station 2

Students will be given a ruler and a protractor. They will measure the angles and lengths of two trapezoids. They will determine if the two trapezoids are similar. Students will derive the relationship between the angles in similar polygons.

Answers

1. $m\angle A = 56$, $m\angle B = 56$, $m\angle C = 124$, and $m\angle D = 124$
2. $m\angle M = 124$, $m\angle N = 124$, $m\angle O = 56$, and $m\angle P = 56$
3. no
4. The base = 3 inches, the sides = 1.3 inches each, and the top = 1.7 inches.
5. The base = 1.5 inches, the sides = 0.65 inches each, and the top = 0.85 inches.
6. 2/1
7. Yes. 3/1.5 = 1.7/0.85 = 1.3/0.65
8. Answers will vary.
9. Answers will vary.
10. Corresponding angles are equal.

Station 3

Students will be given graph paper, a blue marker, a red marker, and a green marker. They will create similar rectangles using graph paper. Students will find the scale factors of the rectangles by physically placing them on top of each other. They will find the scale factors of other rectangles.

Answers

1. Student drawings should depict a blue rectangle that is 12 units long by 9 units wide.
2. Student drawings should depict a red rectangle that is 4 units long by 3 units wide.
3. 3; 3; 1/3; 1/3
4. 3; 1/3
5. 60 units long, 45 units wide
6. 2 units long, 1.5 units wide

7. No, because the rectangles aren't similar. No, because the rectangles aren't similar.
8. Answers will vary. Possible answer: Scale drawings used in architecture.

Station 4

Students will be given a ruler and a protractor. They will find the missing side lengths of similar triangles. Students will discuss the strategy they used to determine whether triangles were similar or congruent. Students will explain why right triangles are not similar to obtuse triangles. Then they will explain why not all right triangles are similar.

Answers

1. $x = 16$, similar
2. $x = 6$, similar
3. $x = 3$, congruent
4. Answers will vary.
5. Answers will vary.
6. no
7. no

Materials List/Setup

Station 1 graph paper; ruler; red marker; green marker; blue marker

Station 2 ruler; protractor

Station 3 graph paper; scissors; blue marker; red marker; green marker

Station 4 ruler; protractor

Discussion Guide

To support students in reflecting on the activities and to gather some formative information about student learning, use the following prompts to facilitate a class discussion to "debrief" the station activities.

Prompts/Questions

1. What is the relationship between corresponding sides in similar polygons?
2. What is the relationship between corresponding angles in similar polygons?
3. What does it mean to scale up a figure by a factor of 4?
4. What does it mean to scale down a figure by a factor of 1/2?
5. How can you find the missing side length in similar triangles?

Think, Pair, Share

Have students jot down their own responses to questions, then discuss with a partner (who was not in their station group), and then discuss as a whole class.

Suggested Appropriate Responses

1. All corresponding sides are proportional.
2. All corresponding angles are congruent.
3. Make the figure 4 times the size of the original.
4. Make the figure 1/2 the size of the original.
5. Use proportions of corresponding sides.

Possible Misunderstandings/Mistakes

- Not correctly identifying corresponding sides of similar polygons
- Not setting up proportions correctly between similar polygons
- Not multiplying scale factor by all sides in the polygon
- Not realizing that not all right triangles are similar because they can have different corresponding angles

Station 1

At this station, you will find graph paper, a ruler, a red marker, a green marker, and a blue marker. Work as a group to construct the polygons and answer the questions.

1. On your graph paper, use the red marker to construct a rectangle that has vertices (1, 1), (5, 1), (1, 9), and (5, 9).

2. On your graph paper, use the green marker to construct a rectangle that has vertices (1, –1), (3, –1), (1, –5), and (3, –5).

3. On your graph paper, use the blue marker to construct a rectangle that has vertices (–1, 1), (–1, 10), (–11, 10), and (–11, 1).

4. What is the ratio between the shorter side of the red rectangle and the shorter side of the green rectangle?

   What is the ratio between the longer side of the red rectangle and the longer side of the green rectangle?

   What is the proportion between the shorter side and the longer side of the red and green triangles?

5. What is the ratio between the shorter side of the red rectangle and shorter side of the blue rectangle?

   What is the ratio between the longer side of the red rectangle and the longer side of the blue rectangle?

   What is the proportion between the shorter side and the longer side of the red and blue rectangles?

continued

6. Which of the rectangles are similar? Write your answer as a proportion. Justify your answer by measuring and comparing the side lengths of each rectangle. Show your work and answer in the space below.

7. What is the perimeter of each rectangle? Show your work and answer in the space below.

8. What is the area of each rectangle? Show your work and answer in the space below.

9. Based on your observations in problems 1–8, what is the relationship between the perimeter and corresponding sides of similar rectangles?

10. Based on your observations in problems 1–8, what is the relationship between the area and the corresponding sides of similar rectangles?

Station 2

At this station, you will find a ruler and a protractor. Work as a group to answer the questions.

For problems 1–2, find the measure of each angle.

1.

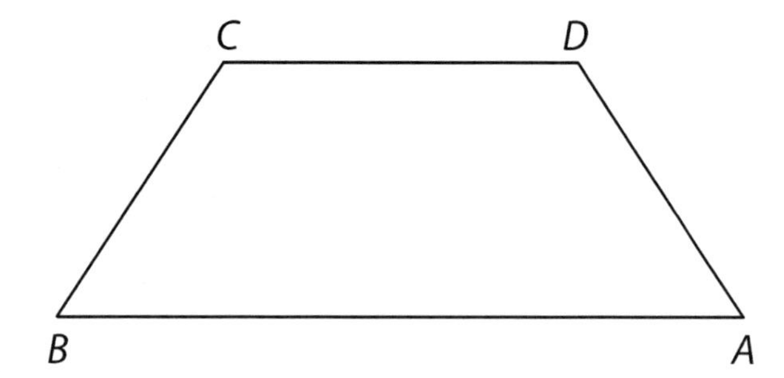

$m\angle A$ = ______ $m\angle B$ = ______ $m\angle C$ = ______ $m\angle D$ = ______

2.

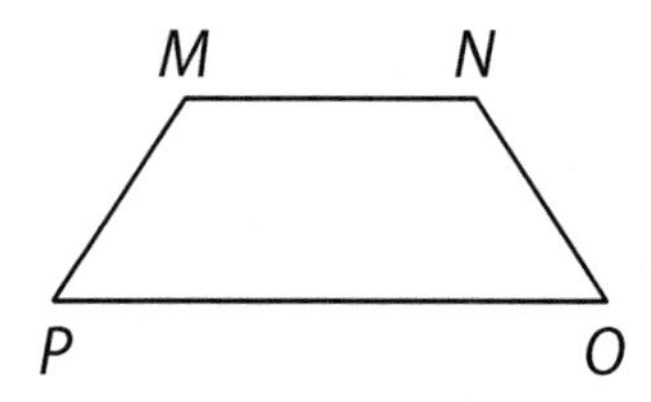

$m\angle M$ = ______ $m\angle N$ = ______ $m\angle O$ = ______ $m\angle P$ = ______

3. Can you say that the trapezoids in problems 1 and 2 are similar based solely on the measure of the angles? Why or why not?

4. What is the length of each side in problem 1?

5. What is the length of each side in problem 2?

continued

6. What is the ratio of the corresponding sides between problems 1 and 2?

7. Are the trapezoids in problems 1 and 2 similar? Explain your answer by setting up a proportion.

8. In the space below, construct and label a trapezoid that is similar to the trapezoid in problem 1.

9. In the space below, construct and label a trapezoid that is NOT similar to the trapezoid in problem 1.

10. What is the relationship between the angles in similar trapezoids and other similar polygons?

Station 3

At this station, you will find graph paper, scissors, a blue marker, a red marker, and a green marker. Work as a group to answer the questions.

1. On your graph paper, construct a rectangle that is 12 units long by 9 units wide. Shade in the rectangle with your blue marker.

2. On your graph paper, construct a rectangle that is 4 units long by 3 units wide. Shade in the rectangle with your red marker.

3. Use your scissors to cut out each rectangle from the graph paper. Place the red rectangle on the blue rectangle.

   How many red rectangles can you fit along the length of the blue rectangle?

   How many red rectangles can you fit along the width of the blue rectangle?

   Place the blue rectangle on the red rectangle.

   What fraction of the blue rectangle can you fit along the length of the red rectangle?

   What fraction of the blue rectangle can you fit along the width of the red rectangle?

4. In problem 3, you found the scale factors of the two rectangles.

   The blue rectangle was larger than the red rectangle by what scale factor?

   The red rectangle was smaller than the blue rectangle by what scale factor?

continued

5. What would be the length and width of a rectangle that has the dimensions of the blue rectangle scaled up by a factor of 5?

6. What would be the length and width of a rectangle that has the dimensions of the red rectangle scaled down by a factor of $\frac{1}{2}$?

7. On your graph paper, construct a rectangle that is 10 units long by 2 units wide. Shade in the rectangle with your green marker.

   Use your scissors to cut this rectangle from the graph paper.

   Can you find the scale factor between the green rectangle and the blue rectangle? Why or why not?

   Can you find the scale factor between the green rectangle and the red rectangle? Why or why not?

8. What is an example of a real-world application of scale factors?

Station 4

At this station, you will find a ruler and a protractor. Work as a group to answer the questions.

For problems 1–3, the triangles are similar. Find the missing length, *x*. Then use your ruler to measure the sides and check the ratios of corresponding sides to verify that the triangles are similar. (Drawings are not to scale.)

1.

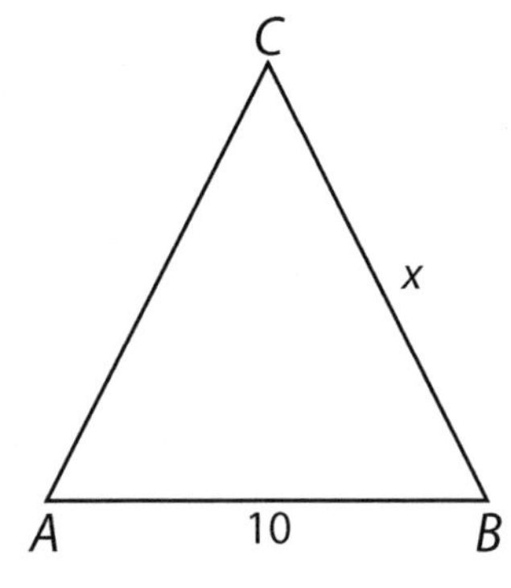

2.

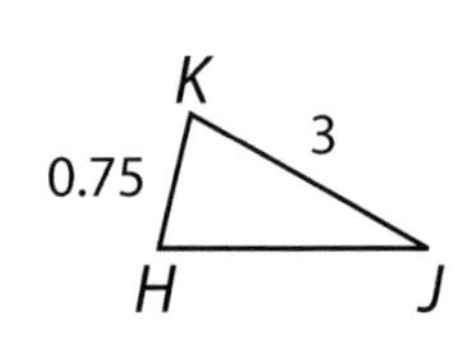

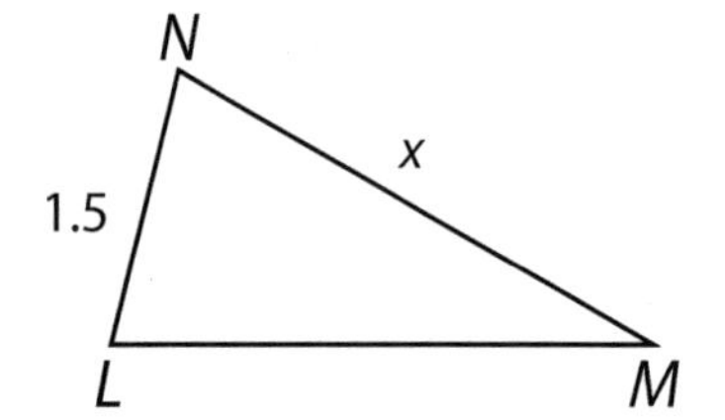

3.

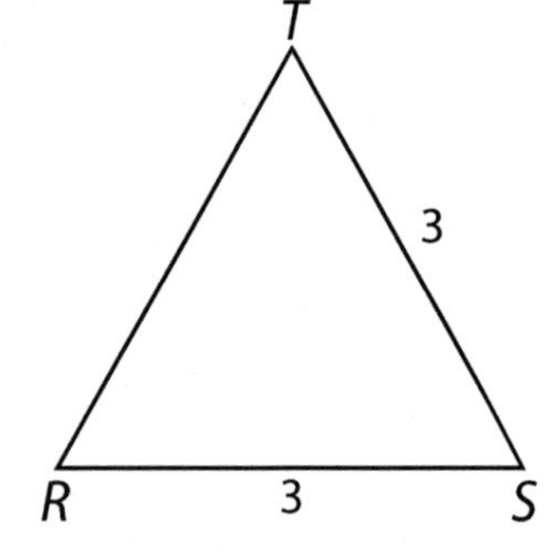

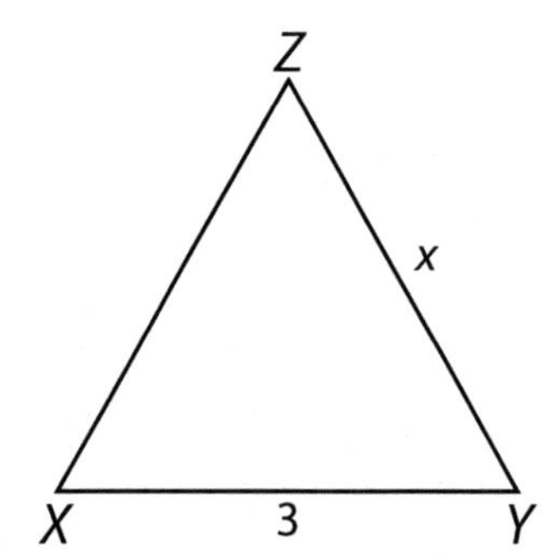

continued

4. What strategy did you use to determine the missing side length, x, in each of the pairs of triangles on the previous page?

5. What strategy did you use to determine whether the triangles were congruent or similar?

6. Is a right triangle similar to an obtuse triangle? Use your ruler and protractor to help explain your answer in the space below.

7. Is a right triangle similar to all right triangles? Use your ruler and protractor to help explain your answer in the space below.

Similarity, Right Triangles, and Trigonometry

Set 2: Sine, Cosine, and Tangent Ratios, and Angles of Elevation and Depression

Instruction

Goal: To provide opportunities for students to develop concepts and skills related to trigonometric ratios for right triangles and angles of elevation and depression

Common Core State Standards

G–SRT.6 Understand that by similarity, side ratios in right triangles are properties of the angles in the triangle, leading to definitions of trigonometric ratios for acute angles.

G–SRT.7 Explain and use the relationship between the sine and cosine of complementary angles.

G–SRT.8 Use trigonometric ratios and the Pythagorean Theorem to solve right triangles in applied problems.★

Student Activities Overview and Answer Key

Station 1

Students will be given a graphing calculator, a ruler, and a protractor. Students will find the sine, cosine, and tangent ratios in a right triangle. Then they will find the measure of each angle in the triangle using these trigonometric ratios.

Answers

1. $\frac{3}{5}$
2. $\frac{4}{5}$
3. $\frac{3}{4}$
4. $\frac{4}{5}$
5. $\frac{3}{5}$
6. $\frac{4}{3}$
7. $\alpha = 36.9°$; ß $= 53.1°$

Station 2

Students will be given a graphing calculator. Students will work together to find the length of a missing side given the length of one side and an angle. Then they will use the trigonometric ratios to find the length of the missing side.

Answers

1. $\sin 50° = \dfrac{x}{12}$
2. $\sin 50° = x/12$; $0.766 = x/12$; $x = 9.19$
3. $\sin 45° = \dfrac{1}{\sqrt{2}}$
4. $x = 12.85$
5. $x = 20.31$
6. $x = 2.13$

Station 3

Students will be given three drinking straws, a graphing calculator, scissors, tape, and a ruler. Students will construct a right triangle out of the drinking straws. They will use it to model an angle of elevation problem. They will use trigonometric ratios of right triangles to find the angle of elevation.

Answers

1. Answers will vary. Sample answer: 5 and 7 inches
2. Answers will vary. Sample answer: approximately 8.6 inches
3. Answers will vary; leg
4. Answers will vary. Sample answer: 8.6 inches; hypotenuse
5. Answers will vary. Sample answer: 7 inches; leg
6. Answers will vary. Possible method: $\tan(x) = 5/7$; $x = 35.5$

Station 4

Students will be given a tape measure and tape. Students physically model an angle of depression. They use trigonometric ratios of right triangles to find the angle of depression.

Answers

1. Answers will vary. Sample answer:

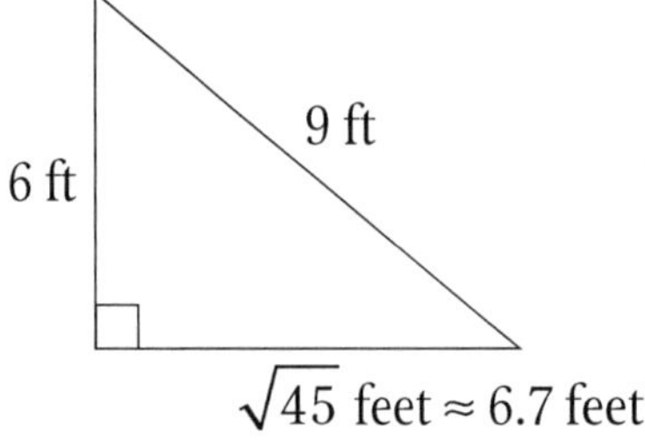

2. Answers will vary. Sample answer: $\cos(x) = 6/9$; $\cos^{-1} = \cos^{-1}(6/9)$; $x \approx 48.2$
3. Answers will vary; Sample answer: It is the angle that is "looking downward."
4. Answers will vary. Sample answer:

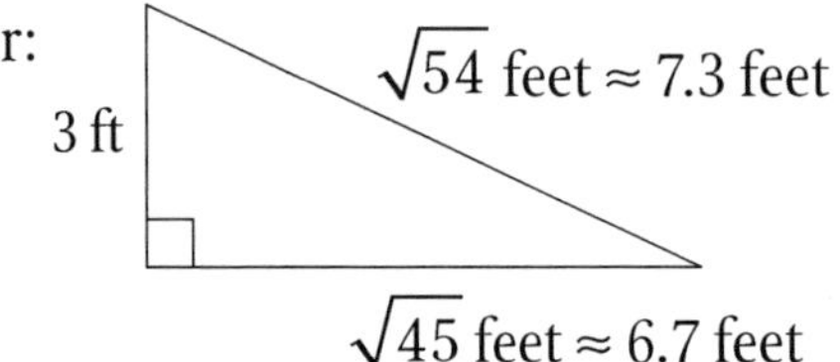

5. Answers will vary. Sample answer: $\cos(x) = 3/7$; $\cos^{-1} = \cos^{-1}(3/7)$; $x \approx 64.6$
6. The angle of depression found in problem 2 is less than the one found in problem 5.

Materials List/Setup

Station 1	graphing calculator, ruler, protractor
Station 2	graphing calculator
Station 3	three drinking straws, graphing calculator, scissors, tape, ruler
Station 4	tape measure, tape

Discussion Guide

To support students in reflecting on the activities and to gather some formative information about student learning, use the following prompts to facilitate a class discussion to "debrief" the station activities.

Prompts/Questions

1. What are the sine, cosine, and tangent trigonometric ratios for right triangles?
2. How do you find the length of a side of a right triangle given one side and an angle?
3. What is the angle of elevation?
4. How do you find the angle of elevation?
5. What is the angle of depression?
6. How do you find the angle of depression?

Think, Pair, Share

Have students jot down their own responses to questions, then discuss with a partner (who was not in their station group), and then discuss as a whole class.

Suggested Appropriate Responses

1. $\sin\theta = \dfrac{\text{opposite}}{\text{hypotenuse}}$; $\cos\theta = \dfrac{\text{adjacent}}{\text{hypotenuse}}$; $\tan\theta = \dfrac{\text{opposite}}{\text{adjacent}}$
2. Use the appropriate trigonometric ratio to write an equation and solve the equation for the missing side.
3. The angle of elevation is the angle between the horizontal and the line of sight of an object above the horizontal.
4. Use the information to draw a right triangle. Then use the appropriate trigonometric ratio to write an equation. Solve the equation for the missing value.
5. The angle of depression is the angle between the horizontal and the line of sight of an object below the horizontal.
6. Use the information to draw a right triangle. Then use the appropriate trigonometric ratio to write an equation. Solve the equation for the missing value.

Possible Misunderstandings/Mistakes

- Incorrectly identifying the opposite side and adjacent side when writing trigonometric ratios
- Incorrectly writing the trigonometric ratios
- Mixing up the angle of elevation with the angle of depression
- Mixing up when to use the trigonometric function and when to use the inverse of the function when finding missing side lengths or angles

Similarity, Right Triangles, and Trigonometry

Set 2: Sine, Cosine, and Tangent Ratios, and Angles of Elevation and Depression

Station 1

You will be given a graphing calculator, a ruler, and a protractor. Work together to answer the questions.

In the space below, construct a right triangle with legs that are 3 inches and 4 inches in length, and a hypotenuse that is 5 inches in length. Label the angle opposite the shorter leg as α (alpha). Label the angle opposite the longer leg as β (beta).

The triangle is a right triangle; you can write trigonometric ratios to represent the values of α and β. Review the trigonometric functions:

$$\sin\theta = \frac{\text{opposite}}{\text{hypotenuse}};\ \cos\theta = \frac{\text{adjacent}}{\text{hypotenuse}};\ \tan\theta = \frac{\text{opposite}}{\text{adjacent}}$$

1. Find $\sin(\alpha)$.
2. Find $\cos(\alpha)$.
3. Find $\tan(\alpha)$.
4. Find $\sin(\beta)$.
5. Find $\cos(\beta)$.
6. Find $\tan(\beta)$.

You can use the trigonometric ratios and your graphing calculator to find the measure of angle α.

7. Find the measures of α and β in the triangle you drew.

Station 2

You will be given a graphing calculator. Work as a group to answer the questions.

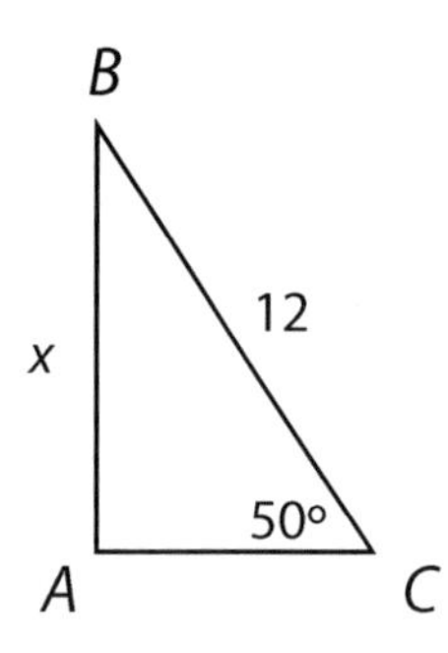

1. How can you write the sine ratio for the 50° angle in the triangle above?

2. Find the value of x using this sine ratio. (*Hint*: First, use your calculator to find sin 50°.) Show your work and answer in the space below.

3. In the space below, construct a 45°–45°–90° triangle. Use your knowledge of the 45°–45°–90° right triangles and trigonometric ratios to prove that $\sin\theta = \frac{\text{opposite}}{\text{hypotenuse}}$.

continued

Similarity, Right Triangles, and Trigonometry

Set 2: Sine, Cosine, and Tangent Ratios, and Angles of Elevation and Depression

For problems 4–6, find the value of x. Round your answer to the nearest hundredth.

4. 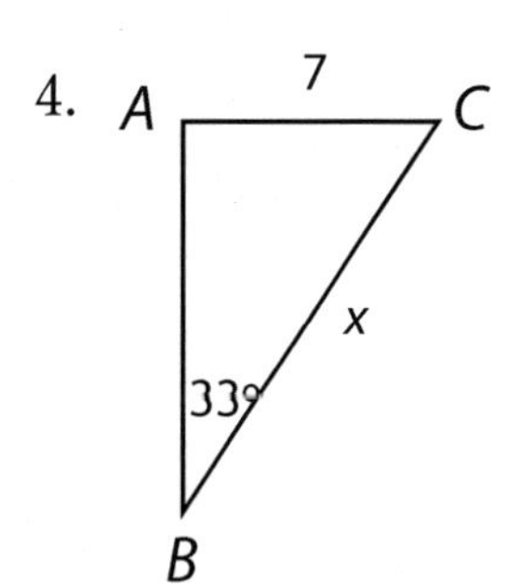

5. 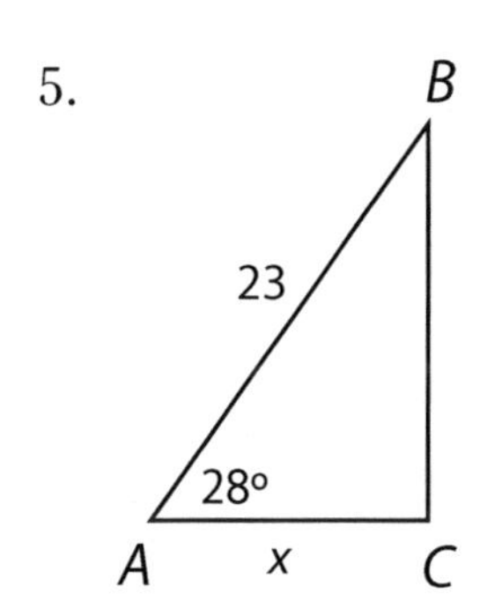

6. 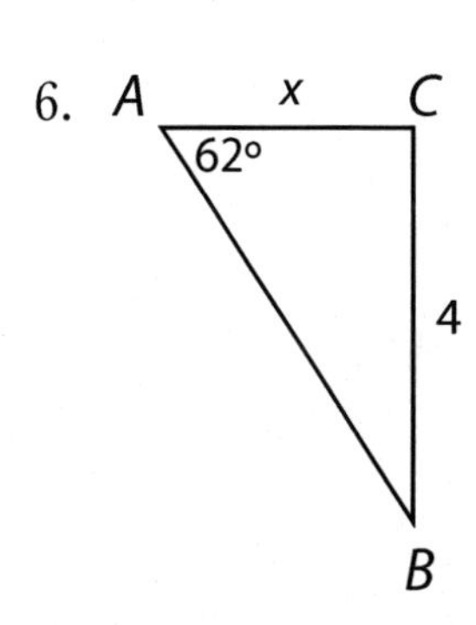

Station 3

You will be given three drinking straws, a graphing calculator, scissors, tape, and a ruler. Work as a group to solve the following problem.

Construct a small right triangle out of the three drinking straws. Use the tape to secure the straws together. Then trim off the excess straws not in the right triangle.

1. What are the lengths of the legs of the triangle?

2. What is the length of the hypotenuse of the triangle?

Imagine that one of the straws in your triangle is a tree. This tree makes a 90° angle with the ground. The tree casts a shadow on the ground. Tape your triangle to your table so it remains upright to represent this situation.

3. What is the height of the straw that represents your tree? Is this a leg or hypotenuse of the triangle? Explain your answer.

4. What is the length of the straw that represents the shadow cast from the top of the tree to the ground? Is this a leg or hypotenuse of the triangle? Explain your answer.

5. What is the length of the straw that represents the distance between the tree and the point on the ground where the shadow ends? Is this a leg or hypotenuse of the triangle? Explain your answer.

6. What is the angle of elevation from the end of the shadow to the top of the tree? Show your work and answer on a separate sheet of paper. (*Hint*: Use trigonometric ratios.)

Station 4

You will be given a tape measure and tape. Work together to solve the following problem.

Have two students stand 6 feet away from each other. Have the first student stand on a chair. Model the diagram below.

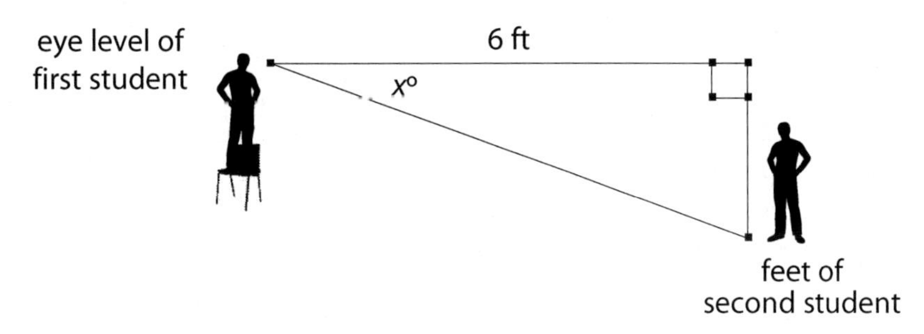

1. Measure the distance between the eye level of the first student and the feet of the second student. What is this distance?

2. Find the angle of depression, represented by x in the diagram, using trigonometric ratios. Show your work and answer in the space below.

3. The angle represented by x in the diagram represents the angle of depression. Why do you think the angle has this name?

continued

Repeat the procedure used in problems 1 and 2 by having two students stand 3 feet apart.

4. Measure the distance between the eye level of the first student and the feet of the second student. What is this distance?

5. Find the angle of depression, represented by x in the diagram, using trigonometric ratios. Show your work and answer in the space below.

6. How does your answer in problem 2 compare with your answer in problem 5?

Circles

Set 1: Special Segments, Angle Measurements, and Equations of Circles

Goal: To provide opportunities for students to develop concepts and skills related to special segments, secants, tangents, angle measurements, and equations of circles

Common Core State Standards

G–C.1 Prove that all circles are similar.

G–C.2 Identify and describe relationships among inscribed angles, radii, and chords. *Include the relationship between central, inscribed, and circumscribed angles; inscribed angles on a diameter are right angles; the radius of a circle is perpendicular to the tangent where the radius intersects the circle.*

G–C.4 (+) Construct a tangent line from a point outside a given circle to the circle.

G–C.5 Derive using similarity the fact that the length of the arc intercepted by an angle is proportional to the radius, and define the radian measure of the angle as the constant of proportionality; derive the formula for the area of a sector.

Student Activities Overview and Answer Key

Station 1

Students will be given a ruler and calculator. They will work together to construct secants that intersect outside a circle. They will derive the relationship between the secants using the secant segment and external portions.

Answers

1. The secants intersect at a point outside the circle.
2. $FA \approx 1.2$ inches
3. $FB \approx 2.9$ inches
4. $FC \approx 1.3$ inches
5. $FD \approx 2.7$ inches
6. $FA \times FB \approx 3.5$ square inches
7. $FC \times FD \approx 3.5$ square inches
8. If two secants intersect outside a circle, then the product of the secant segment with its external portion equals the product of the other secant segment with its external portion.

Station 2

Students will be given a ruler, a compass, a protractor, and a calculator. Students will construct a secant and tangent on a circle. They will measure the angle created by the secant and tangent from a point drawn outside the circle. They will find the measure of the intercepted arcs. Then they will compare the angle and measure of intercepted arcs.

Answers

1.

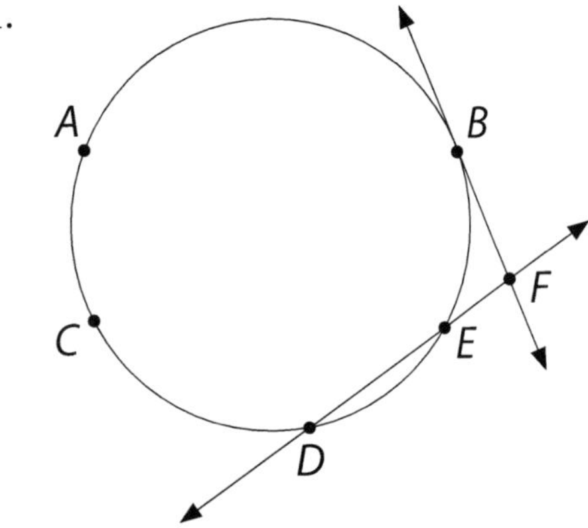

2. 105°
3. 260°
4. 50°
5. The angle formed equals half the difference of the intercepted arcs.

Station 3

Students will be given white paper, a ruler, a compass, a protractor, and a calculator. Students will construct a circle and tangents that intersect at a point outside the circle. They will derive the relationship between the angle formed by the tangents and the intercepted arcs.

Answers

1. Answers will vary. Sample answer: outside the circle

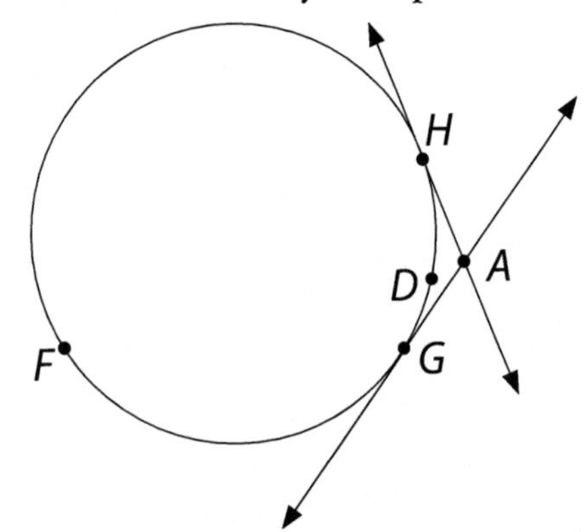

2. Answers will vary. Sample answer: 115°
3. Answers will vary. Possible answer based on answer to question 2: 295°
4. Answers will vary. Possible answer based on answer to question 2: 65°
5. The measure of an angle formed by two tangents drawn from a point outside the circle is half the difference of the intercepted arcs.

Station 4

Students will be given graph paper, a ruler, and a compass. Students will construct circles in the coordinate plane. They will find the equation of a circle given the center and radius, and state the center and radius of the circle when the equation of the circle is given in center-radius form.

Answers

1. $h = 0$; $k = 0$; $x^2 + y^2 = 25$
2. $h = 10$; $k = -4$; $(x - 10)^2 + (y + 4)^2 = 25$
3. Answers will vary. Possible answer: The equation of the circle, $x^2 + y^2 = r^2$, is based on a circle with a center at the origin (0, 0). Both h and k give the center of the circle in relationship to the center of the circle at $x^2 + y^2 = r^2$.
4. (–3, 6); $r = 5$; Sample graph:

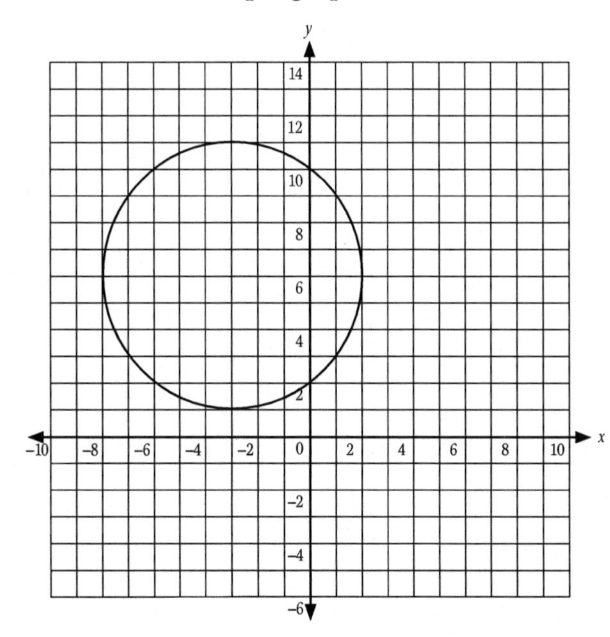

5. (1, 2); $r = 6$; Sample graph:

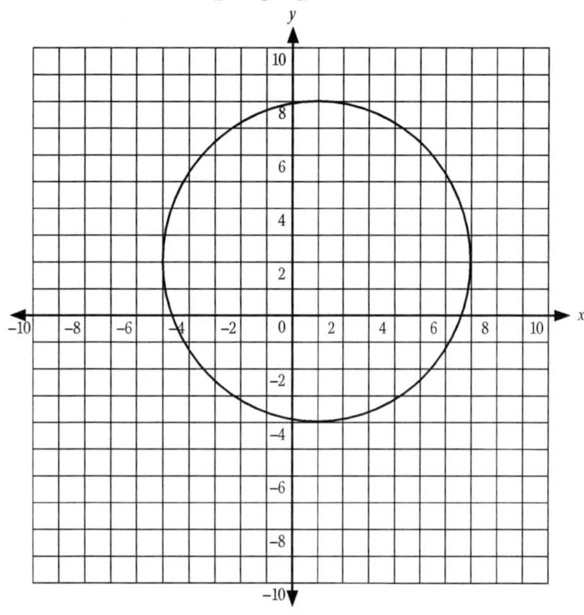

Materials List/Setup

Station 1 ruler; calculator

Station 2 ruler; compass; protractor; calculator

Station 3 white paper; ruler; compass; protractor; calculator

Station 4 graph paper; ruler; compass

Discussion Guide

To support students in reflecting on the activities and to gather formative information about student learning, use the following prompts to facilitate a class discussion to "debrief" the station activities.

Prompts/Questions

1. What is the relationship between two secants that intersect outside of a circle?
2. What is the relationship between the measure of the angle created by the secant and tangent drawn from a point outside the circle and the intercepted arcs?
3. What is the relationship between the measure of the angle created by two tangents drawn from a point outside the circle and the intercepted arcs?
4. How do you write the equation for a circle given the coordinates of the center point and radius?

Think, Pair, Share

Have students jot down their own responses to questions, then discuss with a partner (who was not in the same station group), and then discuss as a whole class.

Suggested Appropriate Responses

1. If two secants intersect outside a circle, then the product of the secant segment with its external portion equals the product of the other secant segment with its external portion.
2. The angle formed equals half the difference of the intercepted arcs.
3. The measure of an angle formed by two tangents drawn from a point outside the circle is half the difference of the intercepted arcs.
4. Use $(x-h)^2+(y-k)^2=r^2$.

Possible Misunderstandings/Mistakes

- Not measuring the correct segments when comparing secants that intersect at a point outside the circle
- Not calculating the measure of intercepted arcs correctly
- Not identifying the intercepted arcs correctly
- Not changing the subtraction sign to a plus sign when a coordinate of the center of the circle is negative

NAME:

Circles

Set 1: Special Segments, Angle Measurements, and Equations of Circles

Station 1

You will be given a ruler and calculator. Work as a group to answer the questions.

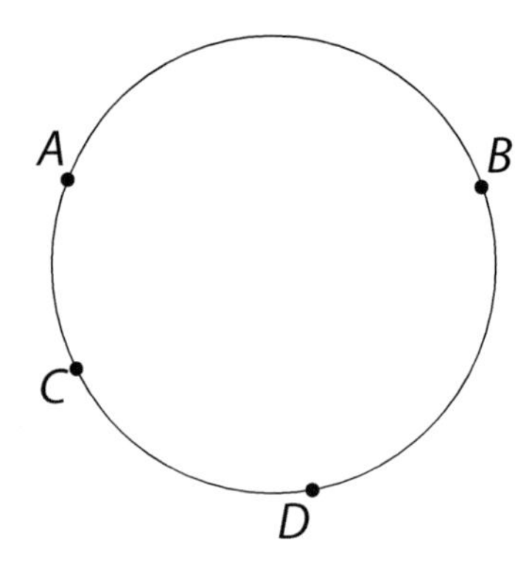

On the circle above, construct a secant through points C and D. Construct a secant through points A and B.

1. Where do the secants intersect?

   Label this intersection as point F.

2. What is the length of $\overline{FA}$?

3. What is the length of $\overline{FB}$?

4. What is the length of $\overline{FC}$?

5. What is the length of $\overline{FD}$?

6. What is $\overline{FA} \times \overline{FB}$?

7. What is $\overline{FC} \times \overline{FD}$?

8. Based on your observations in problems 1–7, what is the relationship between two secants that intersect outside of a circle?

Station 2

You will be given a ruler, a compass, a protractor, and a calculator. Work as a group to answer the questions.

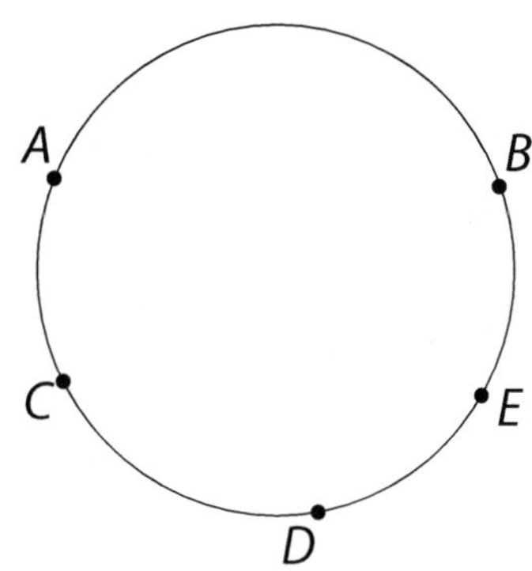

On the circle above, construct a secant through points *D* and *E*. Construct a tangent through point *B*.

1. Do the secant and tangent intersect inside or outside the circle?

2. Label the angle created by the secant and tangent as $\angle F$.

   What is the measure of $\angle F$?

3. What is $m\overset{\frown}{DCB}$? Show your work in the space below.

continued

4. What is $m\overset{\frown}{BE}$? Show your work in the space below.

5. Based on your observations in problems 1–4, what is the relationship between the measure of the angle created by the secant and tangent drawn from a point outside the circle and the intercepted arcs?

Station 3

You will be given white paper, a ruler, a compass, a protractor, and a calculator. Work as a group to construct the circles and answer the questions.

- Construct a circle on the white paper.
- Construct two points on the right edge of the circle. Label the points as *H* and *G*.
- Construct a point on the right edge of the circle between points *H* and *G*. Label this point *D*.
- Construct a point on the left edge of the circle between points *H* and *G*. Label this point *F*.
- Construct tangent lines through points *H* and *G*.

1. Where do the tangent lines intersect?

   Label this as point *A*.

2. What is the measure of $\angle A$?

3. What is $m\overset{\frown}{HFG}$? Show your work and answer in the space below.

4. What is $m\overset{\frown}{GDH}$? Show your work and answer in the space below.

5. What is the relationship between the measure of the angle formed by the point of intersection of the two tangent lines and the intercepted arcs? Show your work and answer in the space below.

Station 4

You will be given graph paper, a ruler, and a compass. Work as a group to construct the circles and answer the questions.

The equation of a circle is $(x-h)^2+(y-k)^2=r^2$.

1. On your graph paper, construct a circle with its center at (0, 0) and a radius of 5 units.

   What is the value of h for this circle?

   What is the value of k for this circle?

   What is the equation of this circle?

2. On your graph paper, shift the circle in problem 1 down 4 units and right 10 units.

   What is the value of h for this circle?

   What is the value of k for this circle?

   What is the equation of this circle?

continued

3. Why do you think h and k are used in the equation of a circle?

4. A circle has the equation $(x + 3)^2 + (y - 6)^2 = 25$.

   What is the center of the circle? Explain your answer.

   What is the radius of the circle? Explain your answer.

   Graph the circle on your graph paper to justify your answers.

5. A circle has the equation $(x - 1)^2 + (y - 2)^2 = 36$.

   What is the center of the circle? Explain your answer.

   What is the radius of the circle? Explain your answer.

   Graph the circle on your graph paper to justify your answers.

Circles

Set 2: Circumference, Angles, Arcs, Chords, and Inscribed Angles

Instruction

Goal: To provide opportunities for students to develop concepts and skills related to circumference, arc length, central angles, chords, and inscribed angles

Common Core State Standards

G–C.2 Identify and describe relationships among inscribed angles, radii, and chords. *Include the relationship between central, inscribed, and circumscribed angles; inscribed angles on a diameter are right angles; the radius of a circle is perpendicular to the tangent where the radius intersects the circle.*

G–C.5 Derive using similarity the fact that the length of the arc intercepted by an angle is proportional to the radius, and define the radian measure of the angle as the constant of proportionality; derive the formula for the area of a sector.

Student Activities Overview and Answer Key

Station 1

Students will be given a plastic coffee can lid, a tape measure, a black marker, a compass, a ruler, and white paper. Students will measure the radius, diameter, and circumference of the coffee can lid. They will derive the relationships between circumference and radius and between circumference and diameter. Then they will solve a real-world problem using circumference.

Answers

1. Circumference; answers will vary.
2. Verify the circle on students' papers.
3. Answers will vary; answers will vary; $d = 2r$
4. the ratio of the circumference of a circle to its diameter, which may be approximated by 3.14
5. Answers will vary; yes, this calculation finds the circumference.
6. Answers will vary; yes, this calculation finds the circumference.
7. Yes, this calculation finds the circumference.
8. $C = 2\pi r$; $C = \pi d$
9. $C = \pi d = \pi(20) = 62.8$ ft
10. $C = 2\pi r$; $6.28 = 2\pi r$; $6.28 = 6.28r$; $r = 1$ inch

Station 2

Students will be given a compass, a protractor, a red marker, a calculator, and a ruler. They will construct a circle, radii, a central angle, and an arc. Students will derive the relationship between the arc and the circumference. Then they will find the length of the arc.

Answers

1. 6.28 inches
2. $\frac{1}{4}$; $\frac{90^\circ}{360^\circ} = \frac{1}{4}$
3. $\frac{1}{4}(6.28) = 1.57$ inches
4. length of arc $= \left(\frac{\text{central angle}}{360^\circ}\right)(\text{circumference})$
5. 2.62 cm
6. 8.85 in
7. 2.36 m
8. 38.47 ft

Station 3

Students will be given white computer paper, a compass, a ruler, a red marker, a blue marker, and a calculator. Students will construct circles, radii, and chords. They will derive the relationship between chords, arcs, and the triangles created by chords and radii. Then they will solve a real-world problem.

Answers

1. 9.42 in
2. 1.05 in; 1.05 in
3. 1.05 in; 1.05 in
4. length of arc = length of chord
5. $\triangle PAB$ and $\triangle PCD$ are congruent because of the SAS theorem. Both triangles have congruent sides (the radii) and the included corresponding angles are congruent since they each measure 40 degrees.

6. $m\overset{\frown}{AB}$ and $m\overset{\frown}{CD}$ are congruent because the central angles are the same measure.
7. $C = \pi d = \pi(10) = 31.4$ in. Cut the pie into 6 slices by dividing 31.4 by 6 to find the arc length of one slice: $31.4/6 = 5.23$ in. Chord length is equal to arc length.

Station 4

Students will be given white computer paper, a compass, a protractor, a ruler, and a calculator. They will construct circles, inscribed angles, and central angles. Students will find the degree measure of the inscribed angle and its intercepted arc. Then they will derive the relationship between the inscribed angle and its intercepted arc.

Answers

1. Answers will vary.
2. Answers will vary. (Should be twice the answer in problem 1.)
3. Answers will vary.
4. Answers will vary. (Should be twice the answer in problem 3.)
5. The measure of an inscribed angle is equal to one-half the degree measure of its intercepted arc.

Materials List/Setup

Station 1 plastic coffee can lid; tape measure; black marker; compass; ruler; white paper

Station 2 compass; protractor; red marker; calculator; ruler

Station 3 white paper; compass; ruler; red marker; blue marker; calculator

Station 4 white paper; compass; protractor; ruler; calculator

Discussion Guide

To support students in reflecting on the activities and to gather some formative information about student learning, use the following prompts to facilitate a class discussion to "debrief" the station activities.

Prompts/Questions

1. How do you find the circumference of a circle given the radius? Given the diameter?
2. In a circle, how do you find the length of an arc given its central angle?
3. In a circle, what is the relationship between arcs and chords?
4. In a circle, what is the relationship between an inscribed angle and its intercepted arc?

Think, Pair, Share

Have students jot down their own responses to questions, then discuss with a partner (who was not in the same station group), and then discuss as a whole class.

Suggested Appropriate Responses

1. $C = 2\pi r$; $C = \pi d$
2. $\text{length of arc} = \left(\frac{\text{central angle}}{360^\circ}\right)(\text{circumference})$
3. length of arc = length of chord
4. The measure of an inscribed angle is equal to one-half the degree measure of its intercepted arc.

Possible Misunderstandings/Mistakes

- Not understanding the difference between a radius and a chord
- Not finding the measure of the correct arc
- Not having the vertex of an inscribed angle on the circle

Station 1

You will be given a plastic coffee can lid, a tape measure, a black marker, a compass, a ruler, and white paper.

1. As a group, use the black marker to mark a starting point on your coffee lid.

   Roll the coffee can lid along the tape measure so you can measure the distance around the edge of the coffee can lid.

   What is the mathematical name for this distance?

   What is the distance around the edge of the lid in inches?

   Repeat this measurement three more times to verify your answer.

2. Trace the coffee can lid on the white paper.

   Use the ruler and compass to find the center of the circle.

3. What is the radius of the circle?

   What is the diameter of the circle?

   How does the radius relate to the diameter?

4. What is π?

continued

5. What is π times twice the radius of your circle?

   Does this match your answer in problem 1? Why or why not?

6. What is π times the diameter of your circle?

   Does this match your answer in problem 1? Why or why not?

7. Do your answers for problems 5 and 6 match? Why or why not?

8. Based on your observations in problems 1–7, what is the formula for the circumference of a circle written in terms of the radius?

   What is the formula for the circumference of a circle written in terms of the diameter?

continued

Circles

Set 2: Circumference, Angles, Arcs, Chords, and Inscribed Angles

9. Larry installed a circular pool in his backyard. The pool has a diameter of 20 feet. What is the circumference of the pool? Show your work in the space below.

10. Lisa is running for class president and passed out buttons that each have a circumference of 6.28 inches. What is the radius of each button? Show your work in the space below.

Station 2

You will be given a compass, a protractor, a red marker, a calculator, and a ruler. Work as a group to construct the circles and answer the questions.

1. In the space below, construct a circle with a diameter of 2 inches. Label the center of the circle as point C.

   What is the circumference of the circle?

continued

2. On the circle, construct a horizontal radius. Use the protractor to create a vertical radius that creates a 90° angle with your horizontal radius.

   Use the red marker to highlight the arc of the circle between the endpoints of these radii.

   What fraction of the circle's circumference is this arc?

   What is the ratio of the central angle you created between the two radii and the total angle measure of the circle? Justify your answer.

3. How can you use the circumference of the circle and the ratio of the central angle to the total angle measure of the circle to find the length of the arc? Explain your answer.

4. In general, what method can you use to find the length of an arc given the central angle?

continued

Circles

Set 2: Circumference, Angles, Arcs, Chords, and Inscribed Angles

For problems 5–8, find the length of the arc using the given information.

5. Circle with radius 5 cm and central angle of 30°

6. Circle with diameter 7 inches and central angle of 145°

7. Circle with radius 0.5 meters and central angle of 270°

8. Circle with diameter 14 feet and central angle of 315°

Station 3

You will be given white paper, a compass, a ruler, a red marker, a blue marker, and a calculator. Work as a group to construct the circles and answer the questions.

On the paper, construct a circle that has a diameter of 3 inches. Mark the center of the circle as point *P*.

1. What is the circumference of the circle?

2. Construct two radii, $\overline{PA}$ and $\overline{PB}$, that create a 40° angle. Draw chord $\overline{AB}$.

   What is the length of chord $\overline{AB}$?

   What is the length of $\overset{\frown}{AB}$? (*Hint:* Use $\left(\frac{\text{central angle}}{360°}\right)(\text{circumference})$.)

3. Construct two radii, $\overline{PC}$ and $\overline{PD}$, that also create a 40° angle. Draw chord $\overline{CD}$.

   What is the length of chord $\overline{CD}$?

   What is the length of $\overset{\frown}{CD}$?

continued

4. Based on your observations in problems 1–3, what is the relationship between chords and arcs?

5. What is the relationship between $\triangle PAB$ and $\triangle PCD$? Explain your answer.

6. What is the relationship between $m\overset{\frown}{AB}$ and $m\overset{\frown}{CD}$? Explain your answer.

7. An apple pie 10 inches in diameter is cut into 6 equal size slices. What is the length of the chord for each slice of pie? Show your work and answer in the space below.

Station 4

You will be given white paper, a compass, a protractor, a ruler, and a calculator. Work as a group to construct the circles and answer the questions.

On the white paper, construct a circle with a radius of 0.75 inches.

1. Plot a point on the circle. Label this point *P*.

   Construct two chords, $\overline{PA}$ and $\overline{PB}$, to create an inscribed angle.

   What is the measure of the inscribed angle, $\angle APB$?

2. Label the center of the circle as point *C*.

   Construct two radii, $\overline{CA}$ and $\overline{CB}$.

   In the space below, find $m\overset{\frown}{AB}$.

Circles

Set 2: Circumference, Angles, Arcs, Chords, and Inscribed Angles

On the white paper, construct a new circle with a radius of 2 inches.

3. Plot a point on the circle. Label this point P.

   Construct two chords, $\overline{PA}$ and $\overline{PB}$, to create an inscribed angle.

   What is the measure of the inscribed angle, $\angle APB$?

4. Label the center of the circle as point C.

   Construct two radii, $\overline{CA}$ and $\overline{CB}$.

   In the space below, find $m\widehat{AB}$.

5. Based on your observations in problems 1–4, what is the relationship between the measure of an inscribed angle and its intercepted arc?

Statistics and Probability

Set 1: Probability

Instruction

Goal: To provide opportunities for students to develop concepts and skills related to the counting principle and simple and compound probabilities for both independent and dependent events

Common Core State Standards

S–CP.1 Describe events as subsets of a sample space (the set of outcomes) using characteristics (or categories) of the outcomes, or as unions, intersections, or complements of other events ("or," "and," "not").★

S–CP.2 Understand that two events *A* and *B* are independent if the probability of *A* and *B* occurring together is the product of their probabilities, and use this characterization to determine if they are independent.★

S–CP.3 Understand the conditional probability of *A* given *B* as *P*(*A* and *B*)/*P*(*B*), and interpret independence of *A* and *B* as saying that the conditional probability of *A* given *B* is the same as the probability of *A*, and the conditional probability of *B* given A is the same as the probability of *B*.★

Student Activities Overview and Answer Key

Station 1

Students will be given 3 pieces of red yarn, 3 pieces of blue yarn, and tape. Students will be given five index cards that have the following written on them: "chicken," "tuna," "white," "wheat," and "Italian." Students will use the index cards and yarn to model the counting principle of independent events.

Answers

1. No, because there are three types of bread available.
2. chicken, white bread; chicken, wheat bread; chicken, Italian bread
3. tuna, white bread; tuna, wheat bread; tuna, Italian bread
4. multiplication; (2 types of meat)(3 types of bread) = 6 possibilities
5. (4)(3)(8)(4) = 384 different meals
6. No; they are independent events.

Station 2

Students will be given index cards with the following subjects written on them: "math," "science," "English," "history," "physical education," and "computer lab." Students will arrange the index cards to create a class schedule. They will derive the counting principle for dependent events by analyzing how to create all possible class schedules.

Answers

1. math, science, English, history, physical education, computer lab; 6
2. Answers will vary.
3. 5; 5
4. Answers will vary.
5. 4; 4
6. Answers will vary.
7. 3; 3
8. Answers will vary.
9. 2; 2
10. Answers will vary.
11. 1; 1
12. Answers will vary.
13. Multiply (6)(5)(4)(3)(2)(1) = 720 possible schedules
14. Dependent, because you can only use each class once in the schedule.

Station 3

Students will be given a number cube and fair coin. Students will use the number cube to model simple probability. Then they will use the fair coin to model simple probability. They will explore mutually exclusive events and provide real-world examples.

Answers

1. Answers will vary.
2. equal
3. 6
4. 1, 2, 3, 4, 5, 6
5. $P(5) = 1/6$; $P(6) = 1/6$

6. $P(>4) = P(5) + P(6) = 1/6 + 1/6 = 1/3$; $P(\text{even number}) = 1/2$
7. H, T; 2
8. 1/2
9. $1/2 + 1/2 = 1/4$; they are independent events.
10. casino and carnival games

Station 4

Students will be given a bag of 4 marbles that are red, green, yellow, and blue. They will also be given a fair coin. Students will use the marbles to model independent and dependent events. Then they will use the marbles and fair coin to model compound probability and mutually exclusive events.

Answers

1. 4
2. Answers will vary.
3. $P(\text{color drawn}) = 1/4$
4. red and green
5. red and yellow
6. red and blue
7. green and yellow
8. green and blue
9. yellow and blue
10. Answers will vary.
11. $P(\text{pair of marble colors}) = 1/6$
12. $P(\text{green}) = 1/4$
13. H, T; 1/2
14. $P(\text{T}) = 1/2$
15. independent
16. Green/T; Green/H; Red/T; Red/H; Blue/T; Blue/H; Yellow/T; Yellow/H
17. 8; $P(\text{G/T}) = 1/8$
18. $P(\text{G/T}) = P(\text{G}) \bullet P(\text{T}) = 1/4 \bullet 1/2 = 1/8$

Materials List/Setup

Station 1 3 pieces of red yarn; 3 pieces of blue yarn; tape; 5 index cards that have the following written on them:

"chicken," "tuna," "white," "wheat," "Italian"

Station 2 6 index cards with the following written on them:

"math," "science," "English," "history," "physical education," "computer lab"

Station 3 number cube; fair coin

Station 4 bag of 4 marbles that are red, green, yellow, and blue; fair coin

Discussion Guide

To support students in reflecting on the activities and to gather some formative information about student learning, use the following prompts to facilitate a class discussion to "debrief" the station activities.

Prompts/Questions

1. How do you use the fundamental counting principle to find the number of possibilities of independent events?
2. How do you use the fundamental counting principle to find the number of possibilities of dependent events?
3. How do you find the simple probability of independent events?
4. How do you find the compound probability of independent events?
5. How do casinos and carnivals rely on probability to make money?

Think, Pair, Share

Have students jot down their own responses to questions, then discuss with a partner (who was not in their station group), and then discuss as a whole class.

Suggested Appropriate Responses

1. Multiply together the number of possibilities for each event.
2. Multiply (number of possibilities) times (number of possibilities – 1) and so on.
3. The simple probability of one event equals the number of true outcomes/number of equally likely outcomes. The simple probability of two independent events equals the probability of the first event plus the probability of the second event.
4. The compound probability of two independent events is the probability of the first event multiplied by the probability of the second event.
5. Casinos and carnivals set up games that have the odds in their favor.

Possible Misunderstandings/Mistakes

- Not understanding the difference between independent and dependent events
- Forgetting to subtract 1 from the number of possibilities each time you perform multiplication using the counting principle for dependent events
- Not realizing that the simple probability of two or more independent events is found by adding the individual probabilities
- Not realizing that the compound probability of two or more independent events is found by multiplying each individual probability by the others

Station 1

You will be given 3 pieces of red yarn, 3 pieces of blue yarn, and tape. You will also be given 5 index cards that have the following written on them: "chicken," "tuna," "white," "wheat," and "Italian." Use these materials and the problem scenario below to answer the questions.

> Jon is at the sandwich shop ready to eat lunch, but can't decide what sandwich to order. He wants to get either chicken or tuna. He has narrowed down the type of bread to white, wheat, or Italian.

1. Is Jon choosing between 2 sandwiches? Why or why not?

2. As a group, determine how many different chicken sandwiches he can pick from by taping the 3 pieces of red yarn to the "chicken" index card and each end of the yarn to the "white," "wheat," or "Italian" cards.

   What sandwich options have you created?

3. Repeat this process with the "tuna" index card and the blue yarn.

   What sandwich options have you created?

4. Based on the number of different sandwiches you have created, what operation (+, –, •, or ÷) can you use to find the number of sandwiches Jon can choose from? Explain your answer.

continued

Use your observations from problems 1–4 to help you answer the next question.

5. A local restaurant offers a four-course meal that includes soup, salad, an entrée, and a dessert for $19.95. You can choose from 4 soups, 3 salads, 8 entrées, and 4 desserts.

   How many different four-course meals can you create? Show your work.

6. Does your choice of entrée depend on your choice of salad? What does this tell you about what type of events these are?

Station 2

You will be given 6 index cards with the following written on them:

"math," "science," "English," "history," "physical education," "computer lab"

Use these index cards and the problem scenario below to answer the questions that follow.

The computer at school is creating a class schedule for a student named Elena. There are 6 class periods in the day.

1. List the possible classes Elena could have first period. How many possible classes did you list?

2. Select one of the index cards as Elena's first-period class. Place the index card underneath this paper.

   What class did you choose?

3. Look at your index cards. How many index cards are left?

   This means there are how many choices for Elena's second-period class?

4. Select one of the remaining index cards as Elena's second-period class. Place the index card underneath this paper.

   What class did you choose?

5. Look at your index cards. How many index cards are left?

   This means there are how many choices for Elena's third-period class?

continued

6. Select one of the remaining index cards as Elena's third-period class. Place the index card underneath this paper.

   What class did you choose?

7. Look at your index cards. How many index cards are left?

   This means there are how many choices for Elena's fourth-period class?

8. Select one of the remaining index cards as Elena's fourth-period class. Place the index card underneath this paper.

   What class did you choose?

9. Look at your index cards. How many index cards are left?

   This means there are how many choices for Elena's fifth-period class?

10. Select one of the remaining index cards as Elena's fifth-period class. Place the index card underneath this paper.

    What class did you choose?

11. Look at your index cards. How many index cards are left?

    This means there are how many choices for Elena's sixth-period class?

12. Select the last remaining index card as Elena's sixth-period class. Place the index card underneath this paper.

    What class did you choose?

continued

13. You created one possible class schedule for Elena. How can you use your answers in problems 1, 3, 5, 7, 9, and 11 to find ALL the possible class schedules Elena could have?

14. Were the choices you selected for each class period dependent on your choices for the other class periods? Why or why not?

Station 3

You will be given a number cube and a fair coin. You will use these tools to explore simple probability.

1. As a group, roll the number cube. What number did you roll?

2. Did the number you rolled have less, equal, or more of a chance of being rolled than the other numbers on the number cube? Explain your answer.

3. How many numbers are on the number cube?

4. List these numbers: ___

   In probability, these numbers are known as your sample space or possible outcomes.

5. What is the probability of rolling a 5 on your number cube? Explain your answer.

   What is the probability of rolling a 6 on your number cube? Explain your answer.

6. How can you use your answers in problem 5 to find the probability P of rolling a number > 4 on your number cube?

continued

What is the $P(\text{number} > 4)$?

What is the $P(\text{even number})$?

As a group, examine the coin.

7. What is the sample space of the coin?

   How many possible outcomes can you have when you flip the coin?

8. What is the probability of tossing the coin heads-up?

9. What is the probability of tossing the coin heads-up, then tossing it heads-up again? Explain your answer.

10. What are some examples of simple probability used in the real world?

Station 4

You will be given a bag of 4 marbles that are red, green, yellow, and blue. You will also be given a fair coin. You will use these tools to explore compound probability. Work as a group to answer the questions.

1. Place the marbles in the bag. How many different colored marbles are in the bag?

2. Pick 1 marble from the bag without looking. What color marble did you choose?

3. How can you use your answers from problems 1 and 2 to find the probability of choosing a marble of the color you found in problem 2?

Place the marble back in the bag. What are the possible different pairs of marble colors in the bag? A possible pair has been given for problem 4 as a model.

4. red and green
5. ______________________
6. ______________________
7. ______________________
8. ______________________
9. ______________________

10. Pick 2 marbles from the bag without looking. What pair of marble colors did you choose?

continued

11. How can you use your answers from problems 4–9 to find the probability of choosing the pair of marble colors you found in problem 10? (*Hint*: These are dependent events.)

Place the marbles back in the bag.

12. What is the probability of choosing a green marble?

13. Examine your fair coin. List the possible outcomes that you could get if you tossed the coin.

    What is the probability of each outcome?

14. What is the probability of tossing the coin tails-up?

15. Are these two events independent or dependent on each other? Explain your answer.

16. List the sample space for finding the probability of choosing a green marble AND then tossing a coin tails-up: *P*(green/tail).

17. How many outcomes did you find in problem 16?

    What is *P*(green/tail)?

18. What is a faster way to find the *P*(green/tail) than listing the entire sample space?